REVELATION TO GENESIS

CYCLES OF HISTORY PATTERNED IN SCRIPTURE AND THE STARS

2nd Edition

By Jay Branch

Proverbs 25:2 (NKJV)
It is the glory of God to conceal a matter,
But the glory of a king to search out a matter.

ISBN 979-8-9994924-4-9

Southern Avenue Publishing, Inc.
PO Box 102
Watkins, Colorado 80137
Email: support@SouthernAP.net

Cover image -
Carina Nebula by NASA's James Webb Space Telescope

TABLE OF CONTENTS

Dedication & Acknowledgements

To Carrie, my loving wife and companion.

As this is my first serious non-technical book, that took most of 30 years and retirement to write, there are innumerable people to thank. Some of their names I have forgotten, others are no longer with us. So, to all those whose lives touched mine, in whatever brief way - Thank you. May these pages be an acknowledgement of what you did for me.

My family, past and present:

Carrie, Jodi, Bob, Tanner, Rebekah, Emma, Leah, Justin, Jessica, Alissa, Dale

Testimonials:

It was VERY enlightening, encouraging and DEEP!! What a great work of art, and love!! Tess W

I thought it was beautifully written, and perfectly edited! I'm still thinking about it and considering the implications of some of your thoughts. Eric E

Preface (to the first edition)

Though I rely on the use of many Hebrew and a few Greek words in this text, please don't think this book is primarily for those with an academic or theological background. Hopefully, you find this text very accessible and easy to understand even if you have absolutely no interest or concern for Biblical languages. My purpose is to present an idea I have found hidden in the Biblical text. That idea spans the Bible from the very first word in Genesis to the very last words in Revelation and should be quite obvious once you have completed this reading. And, by-the-way, "Revelation to Genesis" is not a typo!

As the author of this book, which has some dependence on the use of Biblical Hebrew and a little Biblical (Koine) Greek, I think it important to validate my familiarity with these languages. I am not by any means an expert in Biblical languages but I do have more than a passing interest in them. First, I enjoyed a year of Classical Greek while earning my bachelor's degree. My knowledge of Hebrew came much later and much more informally. While doing an in-depth study of Genesis (it took me a bit over 2 years to complete), it became obvious that I was missing some critical insights by not knowing anything about the Biblical Hebrew language and grammar. So, not wanting to go back to college, I availed myself of two college level Hebrew language and grammar courses, both of which are available to the general public on-line or by purchasing the study materials. Those courses are:

- Basics of Biblical Hebrew Grammar, 2nd ed., by Gary D. Pratico and Miles V. Van Pelt, Zondervan 2007. Course materials are available for purchase on-line at -

  https://courses.zondervanacademic.com/basics-of-Biblical-hebrew-1.

  - Dr. Pratico is Senior Professor of Old Testament and Hebrew Language at Gordon-Conwell Theological Seminary in South Hamilton, Massachusetts.
  - Dr. Van Pelt is the Alan Hayes Belcher, Jr. Professor of Old Testament and Biblical Languages at the Reformed Theological Seminary (RTS), Jackson, Mississippi campus.
- Biblical Hebrew: Learning a Sacred Language, by Dr. Michael Carasik; published by The Great Courses, copyright The Teaching Company 2018. A subscription to The Great Courses is required to purchase any of their available offerings. Visit

https://thegreatcourses.com, search for Biblical Hebrew: Learning a Sacred Language.

- Dr. Carasik has taught Hebrew since 1991 and has been an Adjunct Assistant Professor of Biblical Hebrew at the University of Pennsylvania since 2000.

Additionally, readers may find these references, as used in this text, helpful:

New Testament scripture is taken from the New King James Version (NKJV).

Old Testament Hebrew text references are from the Jewish Publication Society (JPS), Hebrew - English TaNaKh, 2nd Ed (1999), unless noted otherwise.

For those diligent enough to do the work of understanding the Biblical languages better, I have included Strong's Exhaustive Concordance of the Bible Hebrew and Greek dictionary reference numbers. Within this text you will see Strong's numbers bracketed like this - [*Strong's#*]. Where you see an "H" prefixing the number (ie. [H7225]), that references a Strong's Hebrew Lexicon citation. Where the prefix is a "G" (ie. [G225]), that references a Strong's Greek Lexicon citation.

For additional language help with specific Hebrew and Greek word definitions, Strong's numbers are referenced in Vine's Complete Expository Dictionary of Old and New Testament Words, Copyright 1984, 1996 by Thomas Nelson Inc.

Where you see '**ff**' after a Bible verse, this means "and following". It may refer to the rest of the Bible chapter, or suggest you continue reading until the topic changes in the text.

Where you see a verse reference with a lowercase letter, usually an 'a' or 'b' immediately following it—an 'a' means the first portion of the verse while a 'b' (or higher in the letter progression) mean the latter portion of the verse (i.e. Genesis 2:7a and 2:7b).

Where you see '**c**' or '**ca**' preceding a date, this means circa or around that date. Exact dates may not necessarily be known or may be in question.

It is the author's desire that people who are not inclined toward Bible reading find this book of enough interest to read the passages cited. For that reason, throughout this entire work, I have spelled the names of the various Biblical books completely. This might seem a bit excessive to some, but I genuinely hope it helps others.

Author's personal note -

Have you ever had a notion that has been swimming around in your head for years and all you have been able to do with it is say to yourself, "I wish I could tell others about my idea"? Of course, it could be something you thought might actually be important. And then, every once in a while, you experience something that gives you a reason to say to yourself, "Yes! That kind of proves my point. I could use that in a book . . if I ever got around to putting my thoughts on paper (or digitally, as the case may be)". That's how this book came about.

I have been studying the Bible since 1972 and I am passionate about Biblical era history. Not only the historical factors expressed in the Bible but also what was going on in the rest of the world as Biblical events were taking place. My wife will testify to all of this. I can't count how many times I have consumed her attention at the dinner table with something relevant to this book that I just came across. I like to think I'm sensitive enough to stop with the Bible history stuff and let it go just before she begins to glaze over. But, then again, occasionally she will come to me with interesting and relevant information she came across in her daily activities. So, I guess she is actually paying attention, which is certainly one of the many things I love about her.

Preface (to the second edition)

While I was not in a particular hurry to write this book originally, I was in something of a hurry to get the unique ideas and connections on paper. Initially, I only intended to get an outline together and then I would try to get someone else with experience and connections to do the actual writing. In the progress of time, after getting the original outline compiled, it seemed less daunting to start the writing process. There was some research still left to be done, but I felt like I had the kernel pretty well under control. So, I started writing. It went well to begin with, things fell into place fairly quickly and before long I had most of the book written. As stated previously, the research was done piecemeal over almost thirty years so I didn't think there was a lot left to be done.

It wasn't until I started doing the initial research for the follow-up books that I realized there was still a lot left to do on this one. While there isn't anything in the original version of this book that is expressly wrong there are a few things that need clarification and expansion. Those are the issues that I address in this second edition.

If you have a strongly held personal perspective on the Biblical events and verses I address herein you will most likely feel that this new perspective is so unorthodox as to be heretical. Further, you will probably find that my use of alternate meanings and translations of certain Bible words and verses are very self serving. If so, please keep in mind, what I am presenting here is a change from not only conventional Biblical interpretation, but also a complete departure from the comfort and familiarity of linear time. My use of alternate translations and meanings will not be random and capricious, but will always fall within the parameters of acceptable usage, though they will always more clearly demonstrate this new vision of cycles as expressed in the whole of the Biblical narrative. As this new view takes shape, you will see how and why some of the alternate word usages are appropriate to the new paradigm.

In conclusion . . sort of!

Is this new paradigm heresy?! Everything is heretical to someone. The important issue is not what is being called heretical, but who is making that declaration. So, with that said, I will make no further effort to dissuade anyone from their feelings and impulsive view of this new perspective.

Keep this in mind—what you will read in the following pages does not change anything about the work of salvation by Jesus on the cross. Nor does it imply anything new about our fallen world and how we are separated from God without the sacrifice that Jesus made for us all.

As I contemplate the implications of this new paradigm of recursive cycles, it might be unavoidable to suppose it allows room for reincarnation. Nothing could be further from the truth. Hebrews 9:27-28 tells us -

27*"And as it is appointed for men to die once, but after this the judgement,* 28*so Christ was offered once to bear the sins of many. To those who eagerly wait for Him He will appear a second time, apart from sin, for salvation."*

The events of Revelation are inevitable and we must be prepared! He is our Savior, and we are only allowed one chance to accept His grace in this present cycle.

PART 1: An Old & New View

Acts 17:10-12

10 Then the brethren immediately sent Paul and Silas away by night to Berea. When they arrived, they went into the synagogue of the Jews. 11 These were more fair-minded . . . in that they received the word with all readiness, and searched the Scriptures daily to find out whether these things were so. 12 Therefore many of them believed. . .

1.1 A perspective on Bible Study

Before any other matters, I want to present my views on Bible study and my own approach to study.

"You have to know what it says before you can know what it means."

This is the essence of study for me. Whether you are studying the Bible, studying to be more informed about your work, more knowledgeable about the political, social or religious environment you are in, it is important to take the time to <u>know</u> what is involved and how you are involved in it. What are the basics, what elements are involved? You shouldn't come to any conclusions before you know the essence of what is being presented.

When studying the Bible it also means not bringing your meaning *to* the Bible, but taking meaning *from* it! Far too often we approach the Bible with a preconceived notion about what we expect it to say. We trust our pastor or our favorite teacher to provide the details for understanding. And, for the most part, that's a good approach. Don't be afraid to tease out the meaning yourself. By taking the time and energy to do the research you are much more likely to remember what you've learned!

If you are inclined to study the Bible, as I have been since 1972, it is important not to be in a hurry. Too often we are consumed by a need to get through the text and finish, so we can get on with other stuff in life or just to have that satisfied feeling of having read more. So, what if something seemed a bit off with the text or the situation people were in—I finished and can now move to the next book or chapter.

I would like to suggest that you — **slow down!** If something seems off, find out why. Those are great opportunities to do more in-depth study. It's the in-depth study of troubling texts that will ultimately, convince you that you can always trust the Bible. Take time to tease out the meaning of key words, phrases or concepts. There are numerous online resources that make this level of study very quick and easy in the internet age.

When I was preparing the study and notes I've used to write this book, those study tools were all hard paper books that required hours of tedious review and reference. My very first in-depth study was Genesis. I had no knowledge of the Hebrew language, so I meticulously looked up almost every word in the first few chapters and many more in the rest of the 50 chapters of Genesis. My purpose was to understand the roots of the text. What did the original writer (Moses) mean when he said something and why did he say it that particular way? You will understand what I mean as you read further. One of the prime motivations for writing this book involves some very important subtleties about the very first word of Genesis (1:1).

I spent another two years studying Daniel. I don't remember what drew me to Daniel, as opposed to just continuing chronologically and working through Exodus, but God had a particular plan and purpose. Daniel revealed a finer level of understanding—that of vision, prophecy and history—that Daniel alone could see.

Upon finishing Daniel it seemed natural to start studying Revelation. I had a year of Classical Greek in college so teasing out the meanings of Greek words was a bit easier than the previous two books written in Hebrew. By the time I was half way through Revelation it was clear to me that Revelation could easily be referred to as Daniel book 2 (or, 2 Daniel, if you will). The imagery, vision and language are so similar as to have been written by the same person. Understanding, of course, Revelation was written more than 600 years following Daniel by John (the Revelator).

Let me take a brief step back and give you a few meaningful examples of what can be gleaned by taking your time to know what the Bible is saying, so you can more fully understand what it means. The first example is addressed more fully in my second book, and the second example is addressed in my third book. (At this time, my second book's working title is "In the Image of God", and my third book's working title is "The Heavens and The Earth".)

Example 1, from Daniel 1: After generations of disobedience to God's will and law, the people of Judah are taken captive by Nebuchadnezzar II and exiled in Babylon. The idea of Babylon is a recurring theme in Revelation as well as the new home of the exiled Jewish people. We may begin to refer to the Israelites as Jews by this time, as the name Jew is a variant of Judah, the chief of the tribe of Israel that possessed the southern half of the Promised Land.

The opening chapter of Daniel tells about Nebuchadnezzar's desire to re-educate and retrain some of the younger new exiles in the ways of the Babylonian culture and the kings expectations. While not disagreeing with the training per se, Daniel and his three Jewish friends—we know them as Shadrach, Meshach and Abed-nego (their Hebrew names are respectively; Hannaniah, Mishael and Azariah)—respectfully refused to defile themselves with the king's idea of food and drink. Daniel's Babylonian name was Belteshazzar, though he seldom refers to himself as such in his own book.

After a challenging beginning to their training, Daniel and friends are able to obey the will and laws of their God while refusing the food and drink of the king. After an agreed upon trial period following their own Kosher food rules, they were (Daniel 1:20) "..found to be ten times better than all the magicians and exorcists throughout his realm." Now, here's the catch—the Hebrew text does not say "ten times better", it says they were "ten[H6235] hands[H3027] better". The only Bible translation I have ever found that says that is Young's Literal Translation, which is a very old word-for-word transliteration from the Hebrew. If our Bible actually said that it would not make a bit of sense. In fact, this is the only place in the Bible where those Hebrew words appear together. So, absent any other context, the translator made the text make sense.

The point of my second book deals with what it means to be a "ten hand"! A ten hand compared to what? One very interesting fact about ancient Babylon (or, through history - Ur, Sumer, Chaldea, and numerous other names) is, it is the place of origin of the 360 degree circle, and more generally, the base 12 math that we see all around us. That base 12 math, as it emerges from Babylon, can be seen in the 12 hour clock, 12 inch foot, 12 month year, 12 zodiacal images, etc. Is there something more understated and hidden in this curious discovery—something having to do with the number *six—or, possibly, a "12 hand"*?

Example 2, from John 4: If it can be said that I have a favorite single chapter in the entire Bible, this would be it. There is so much to glean from this single chapter. However, I want to focus on what Jesus said to the famous woman at the well. We all know of her, but her challenge to Jesus is worth a more in-depth look. First, the unnamed woman is more aware of her situation than most people accept. She knows she lives among a group of people whose lineage has been tainted by the Assyrians.

In 722BC the Assyrians invaded and took captive the "10 northern tribes" of Israel. These would be the "10 lost tribes" that some intrepid historians continue to search for. I contend, not only will they not be found, but they are entirely unfindable. The Assyrians perfected the military strategy of invading a country and hauling off a significant portion of that population to fight their next battles, leaving some of them in place, in a land not their own. This effectively pacified the invaded country by leaving people who were not willing to fight for something that they didn't identify with. The Assyrians left many non-Israelites in the land of northern Israel. The woman knew very well that she was part of this mixed group known as the Samaritans.

She is surprised when Jesus speaks to her—presumably, He is a more orthodox Israelite from the south—though actually He grew up in, and began His ministry from Nazareth, in the Samaritan north, like her. Somewhat unexpectedly, she makes a comment about the differences in their situations. She states that the Samaritans worship "..on this mountain.." (Mt. Gerizim) while acknowledging that good Jews worship in Jerusalem, at the Temple Mount.

Jesus's response to the woman is frequently downplayed but essential to understanding the nature of the God we worship. He tells the woman that a time is coming when the place where someone worships will not be as important as Who they worship. His actual statement is (John 4:23-24):

> 23 *"But the hour is coming, and now is, when the true worshipers will worship the Father in spirit and* ***truth****; for the Father is seeking such to worship Him.*
> *24God is Spirit, and those who worship Him must worship in spirit and* ***truth****."*

An alternative translation for the word truth [G225] in these passages, is reality! Try rereading the passages with the substitution inserted. It presents an interesting dichotomy between Who and What God is, and who and what we are. The only version of the New Testament I have ever found that translated it that way is a somewhat obscure version of the New Testament called the J.B.Phillips Translation.

Truth or Reality? An interesting choice. What I see being offered by Jesus is the distinction between something utterly unlike us—Spirit (πνεύματι) [G4151], while He also recognizes we don't really have a choice. We have to deal with the earthly—reality (ἀληθείᾳ)[G225]. Jesus challenges us to seek both!Diligently and prayerfully searching scripture is a key component to understanding the Spirit that is God.

Grounded in the Promise

Let me conclude these examples with the final lesson from John 4, a lesson I seek to fulfill every day. Beginning in John 4:46 a story is told about a young nobleman who travels to find Jesus, hoping to convince Jesus to follow him home to heal his sick son. There is no real indication the young nobleman is necessarily a Jew as he addresses

Jesus. He tells Jesus of the situation of his dying son. Though Jesus was clearly disappointed that this was another miracle seeker, for He had crowds of them pressing around Him frequently, His response was everything you would expect from Jesus (4:50), "Go your way; your son lives". Hallelujah!!! The young nobleman's response is where the lesson resides. The same verse concludes with, "So the man believed the word that Jesus spoke to him, and he went his way."

He didn't question Jesus's statement—will it be today, will it happen immediately or over time, should I plan on giving him some aspirin or Tylenol and call you in the morning? What should I expect? No, it simply says, he took Jesus at His word, believed, then turned and went home! It isn't always easy to be that strong and grounded in faith, but the blessing follows the promise. Seek that unquestioning belief. You'll know it when you know His Word, the Bible.

The Knowledge!

I am going to challenge almost everyone's sensibility in the coming pages. Whether that be an orthodox theological perspective, or to challenge a comfortable standard model of the natural world. Let us consider an acceptable Biblical framework that allows for such a significant change in our frames of reference as will be presented here. As we learn in Daniel 12:4,

> *"But you, Daniel, shut up the words, and seal the book until the time of the end; many shall run to and fro, and* ***knowledge*** *[H1847] shall increase."*

Then, John the Revelator is told to open the book(s) in Revelation 20:12,

> *"And I saw the dead, small and great, standing before God, and books were opened. And another book was opened, which is the Book of Life. And the dead were judged according to their works, by the things which were written in the books."*

Much of the book of Revelation deals with breaking seals and manifesting judgement on the world.

Again, Hebrew grammar and syntax come to the rescue in understanding that things might change, and change significantly in the end times. The Hebrew word highlighted above is actually "the knowledge", with a definite article present in the original Hebrew text in Daniel. So, the knowledge of what? I'm certain from the context that it is scripture that will be opened and revealed and the specific knowledge of God's Word will be revealed in a new and unexpected way. We may be living the time of that particular increase in the knowledge of God's

Word and His specific meaning that is particular to us and no one else in this current cycle.

Final thoughts before we embark -

My view of the Bible is that it is accurate in all aspects of life that it touches on! As it relates to human nature, it offers us hope; as it relates to history, it offers us pattern; as it relates to science, it gives us much more information than typically thought. Where the Bible deals with science it is always completely accurate. We don't recognize that accuracy because it does not relate science in what we would consider scientific terms, or terminology. This book is intended to be a challenge to established history and science—a complete recontextualization.

The point of this book is to illustrate a subtle truth hidden in the entirety of the Hebrew/Christian Bible. That truth refers to CYCLES, not only of human history, but of planetary and, possibly, universal history.

Cycles - as referred to in this book, are repeating historical sequences in different geological/ cosmological contexts within very different time frames. A cycle very likely has a discrete beginning and end, though a typical cycle covers such long time spans that it is very difficult to recognize those beginning and end points. It is through geological, archaeological, and literary (sometimes oral) contexts and traditions that we are given a glimpse of when a cycle begins or ends.

"Those who forget history are condemned to repeat it," the notable quote of George Santayana from his work The Life of Reason: Reason in Common Sense, reminds us that patterns of history and human behavior are identifiable and should be factored into future considerations. All too often, they are not. This book seeks to expand on Santayana's warning and suggests that there are even more grand, though frequently more obscure aspects than those he was concerned about.

Santayana warns us to watch . . .

The Bible tells us where to look . . if we are willing to see!

1.2 Patterns throughout Time

I think everyone has been challenged by that paradoxical and irritating question, "What if you could live forever, what would you do?" This book isn't going to answer that question. This book deals with *time* in a very unique way—a way I think the Bible has been quietly hiding for thousands of years. It is only now, with new emerging views of the universe that theoretical physicists and mathematicians are offering, that we may now understand time in a way the Bible has always understood it. We may be on the doorstep of forever, and it is not what you think it is.

The Bible has always sought to inform us by telling us stories that repeat over long periods of history (referred to as a '*type*'), or offer repeating paradigms of success or failure, such as -

- The story of Joseph and his success in the Pharaoh's court, compared with the story of Daniel's success, first in Nebuchadnezzar's court and then in succeeding Medo-Persian regimes.

- The Bible repeatedly refers to the Adam of Genesis as the "First Adam", perfect and human, followed by Jesus as the "Second Adam", perfect and divine.

- We also see Moses as a type of Christ Jesus as the deliverer from bondage.

- And, Noah as a type of Christ Jesus as the deliverer into a new creation.

In a secular framework we see a different order of repetition. I would direct you to a very interesting website—"Foundation for the Study of Cycles" (cycles.org)—for a glimpse of how many repeating paradigms have already been identified in human and natural history.

The most important repeating paradigm, as concerns this book, is how often life has been utterly wiped out by asteroids (or meteors) and the poisoning of the oceans (ocean anoxia) by the same. It is commonly accepted that, most recently, the Chicxulub asteroid impact of the Cretaceous period killed the dinosaurs some 65 to 66 million years ago. Compare this, and the following list, with the accounts of Revelation 8—those of meteors falling into and poisoning the oceans (8:8-9); and, of the asteroid called Wormwood (meaning, bitterness) (8:10-11) falling to earth and killing much of land and river life, at the time of the end!

Additionally, consider these extinction events and how similar they are to the events Jesus describes in Matthew 24:29ff and the book of Revelation chapters 6 through 8.

The Chicxulub event and the following four events are called the “Big Five” in Earth’s history and include catastrophic impacts in some cases, but most involve multiple factors like volcanism, climate change, sea-level shifts, ocean anoxia, or atmospheric changes, but all represent extinction level events:

- Ordovician-Silurian extinction (~444 million years ago): ~85% of species lost; main causes: intense glaciation, sea-level fluctuations, and changes in ocean/atmospheric chemistry.

- Late Devonian extinction (~372–359 million years ago): ~75% of species affected (multiple pulses); causes: global cooling, ocean anoxia, and volcanism.

- Permian-Triassic extinction ("The Great Dying," ~252 million years ago): ~96% of marine and ~70% of terrestrial species lost (Earth's worst); primary cause: massive volcanic eruptions releasing greenhouse gases, causing extreme warming, ocean acidification, and anoxia.

- Triassic-Jurassic extinction (~201 million years ago): ~76% of species lost; main cause:

extensive volcanism from Central Atlantic Magmatic Province (linked to Pangaea breakup), leading to climate change and ocean acidification.

As of this writing, no extinction-level asteroids (those larger than about 0.6 miles across capable of global catastrophe) are being tracked with any credible risk of striking Earth. NASA's Center for Near-Earth Object Studies (CNEOS) and its international partners monitor more than 41,000 near-Earth asteroids. Scientists have identified approximately 96% of the largest threats (over 1 kilometer or 0.6 miles), totaling around 878 known objects, none of which pose an impact hazard in the next century. The most watched candidate, asteroid 99942 Apophis (about 1,100 feet across), will pass safely within 20,000 miles of Earth on April 13, 2029 —closer than many satellites—yet carries zero risk of collision for at least the next 100 years. Recent close approaches, such as the bus-sized 2026 EG1 (35 to 50 feet across), passed harmlessly at roughly 197,000 miles.

Smaller "city-killer" asteroids (140 meters to 1 kilometer, or about 460 feet to 0.6 miles) remain a greater near-term concern: an estimated 25,000 exist, but only about 40% have been detected.

These could devastate regions or continents, yet none currently threaten impact.

Governments worldwide demonstrate growing readiness through coordinated planetary defense. The United States leads via NASA's Planetary Defense Coordination Office, building on the successful 2022 DART mission that altered an asteroid's path. The NEO Surveyor telescope, slated for launch soon, aims to meet the congressional goal of cataloging 90% of hazardous objects 140 meters and larger by 2030. Biennial international exercises with FEMA and partners simulate responses ranging from regional evacuations to deflection. The European Space Agency (ESA) contributes missions such as Hera and the planned NEOMIR infrared sentinel. The United Nations Office for Outer Space Affairs coordinates the International Asteroid Warning Network and Space Mission Planning Advisory Group, ensuring global alerts and mitigation planning.

Still, gaps persist. Undetected smaller asteroids could arrive with little warning, and deflection technology, while proven in principle, remains untested against the largest threats. Secular authorities are watching for the next threat from the skies!

In non-Biblical paradigms we can now see some Biblical themes patterned throughout history. We see a certain level of chaos and cataclysm shaking the earth and then life recovers and starts again. According to geology and chemistry, it is inevitable that life will start again. That is just what the Bible tells us will happen.

We will now embark on a journey that, by necessity, begins exactly where it ends. This journey will involve the use and accounting of the idea of time in a very unfamiliar way. I will treat time not as fixed, linear and predictable; but, as flexible, cyclical and predictable.

1.3 A New Perspective on Time

We tend to take time very much for granted—we wake up, go to work, enjoy our family and friends, engage in our side projects, go to bed and wake up again—in an endless repetitive cycle. But could time actually be something much more malleable and flexible?

This book will take a step outside of normally accepted parameters of research and study. Though I will offer information and materials that are researched and verifiable, I offer the following caveats to prepare my reader to step onto a road not yet traveled and a path to a destination that I have not seen anyone else reach.

Albert Einstein and, more recently, Richard Feynman, among numerous other notable theoretical physicists, have come to the conclusion that we live not in a 3-dimensional world of space, but a 4-dimensional world of space-time, meaning, we have to view time as something other than the stuff measured by a sundial, the clock on the wall or our cell phone. Time can stretch or compress depending on speed and proximity to high gravitational objects. It is helpful to understand this aspect of physics in order to understand what follows in this book. The topic of time and space-time will be dealt with more fully in a subsequent book in this collection about heaven.

Albert Einstein was the theoretician who first confused us with the concept of "Relativity". Relativity tells us that time has very little to do with our accounting systems. Those accounting systems are the ones we like to call clocks and calendars. They are the systems that help us order our lives and do things on schedule and represent how we deal with the linear flow of our lives. We are born and we die and there are countless things that happen in between. Relativity tells us that's just a perspective. Time itself, as a physical dimension of the created universe, is something very different.

The late Dr. Stephen Hawking was obsessed with time, as a dimension of the universe, and wrote numerous books on the subject. Posthumously to Dr. Hawking, Thomas Hertog, a colleague, collaborator and doctoral student of Dr. Hawking, in his book "On the Origin of Time: Stephen Hawking's Final Theory", finalized Hawking's notions of how the universe actually works and how time is something very mysterious, indeed! To quote Hertog's book (pg. 190),

> *"By contrast, quantum cosmology inquires about the very origin of the physical environment. It descends all the way to the level of the*

> *quantum observations and, not only that, it strives to do so in the remote realm of the big bang where observership has a hand in how the laws of physics came about. Far from being irrelevant, the intermingling of the ghostly world of superposition is crucial here. It elevates backward-in-time reasoning from a mere retrospective element in the study of this history to a retroactive component that creates this history."*

A lot of theoretical physics gobbledygook to say that at a quantum level—at the elemental level that defines the universe we can know—time has a paradoxical aspect that allows it to create itself. That specific aspect is referred to as "non-locality" and "superposition" (or, quantum entanglement), you may know it more familiarly as - God.

The point I want to emphasize here is that time isn't what we imagine it to be. It could be that repeating history is very possible, and linear time is not as rigid as we experience it to be. We think of time as linear because, in our relatively short life spans, that's how we measure it and that's how we experience it. We use a clock to tell us when history starts but it does not tell us when it will end. We remember the past but not the future. Time, as we experience it, is just a bookkeeping method to account for the past and a way to estimate the future.

A more important question is—does God experience time the same way we experience it? Is the God of "reality" bound to the same bookkeeping methods we use—atomic clocks and history books and such?

In Revelation chapters 6 and 19, where Jesus is seen riding a white horse to deliver justice and make war against the evil that is destroying His people and the planet—do you suppose God is looking at His watch and at a certain time says, "Ok Jesus, time to mount up, you'll be leaving soon."?

I think not!

If God is using any kind of time keeping mechanism, it would be creation itself—the stars and celestial bodies that the Bible describes as being His accounting system. That is where this book is headed.

1.4 A View of Precession

Have you ever looked into the night sky and thought how amazing, vast and unchanging is that sky? If you did, you're not paying enough attention! Even before our current time of satellites and space stations and endless human-produced objects orbiting the earth, the night sky changed inexorably from one night to the next.

To further confuse an already questionable proposition about flexible space-time, I offer the "Precession of the Equinoxes" (precession). The calculation of precession defies common sensibilities, as a precessional cycle takes fully 25,772 years. Yet, there is no shortage of ancient civilizations that seemingly had the precession factored into their calendars.

The Precession of the Equinoxes describes the axial wobble of the earth as it orbits the sun. This wobble is caused by gravitational torques from the Sun and Moon on earth's equatorial bulge, making the axis trace a slow circle in the sky. As calculated in modern times by NASA, a precessional cycle can be recognized by the setting of some stars and their rising 25,772 years later. Plato seems to have been aware of precession. It is often referred to as Plato's Great Year.

Precession may be loosely recognized as the progress of 12 zodiacal cycles of 2160 years, which equals a 25,920 year precession. While it is commonly accepted that precessional cycles can vary a bit, the 148 year difference between NASA's calculation and ancient cultural variation is common.

A precessional cycle progresses counter to the commonly observed zodiacal cycle. While the zodiac is seen to move clockwise, or from east to west in the sky; the precessional cycle moves counterclockwise, or west to east. It is this counterclockwise progression of time and space that this book is based on.

Some of the civilizations that may have recognized precession are -

- The Greeks, Hipparchus (c 190-120 BC) was the first to explicitly document and measure it with reasonable accuracy.

- The Babylonian/Chaldeans (c. 1000-300 BC) Hipparchus relied on their long-term records of equinox/solstice positions and star catalogs. The Babylonians were the first to organize the zodiac and use if for calculations and measure of time and cycles.

- The Egyptians (Dynastic period, c. 3000–30 BC) Often claimed alignments in pyramids like Giza oriented to past pole stars like Thuban, shifts from bull/Taurus to ram/Aries symbolism around 2000 BC, or the Dendera Zodiac, but no direct surviving texts confirm explicit knowledge.

- Sumerians (Mesopotamian precursors, c. 4000–2000 BC) Some alternative interpretations (e.g., animal motifs on artifacts like the Uruk Vase symbolizing solstice/equinox constellations shifting with precession) suggest awareness, but evidence is symbolic and indirect, not quantitative.

- Hindu/Indian traditions (Vedic period and later, c. 1500 BC onward) Some scholars link long yuga cycles or astronomical references to precessional effects, but this is speculative and not a direct recognition of the mechanism.

- Mayans (Mesoamerica, c. 2000 BC–1500 AD) Speculative claims tie the Long Count calendar (~5,125-year cycles) or Pleiades observations to precession, but mainstream Mayanists question direct knowledge.

- Paleolithic/Prehistoric Europeans (e.g., cave art sites like Lascaux, Chauvet, or Lion-Man sculpture, c. 40,000–15,000 BC) Highly controversial archaeoastronomical theories (e.g., animal symbols as zodiacal markers tracking precessional shifts) suggest extremely ancient awareness.

How could a process that takes longer than any known civilization has survived be recognized, let alone, if those ancient cultures were thought to be primitive? Once the process of precession was known, there had to have been a way to trace the long slow precession over unimaginable generations and time periods. Is it possible that such knowledge is passed down not from generation-to-generation, but from cycle-to-cycle as awareness is traced beyond a particular culture? This is a question that should have an answer later in this book.

I would like to acknowledge the work of the late Dr. Carmen Boulter, Egyptologist from the University of Calgary, and her video series entitled "The Pyramid Code". In this series Dr. Boulter finds the ancient Egyptians recognizing grand ages of enlightenment/decline/descent/recovery/back to enlightenment that span tens of thousands of years. How does she characterize those ages (cycles)—the enlightenment is the Golden Age, it declines into the Silver Age, further descends into the Bronze Age, and finally cratering in the Iron Age, only to begin an ascent into new cycle of recovery in a new Bronze/ Silver/Golden Age.

Each of these ages has a sign, a zodiacal marker that characterizes the place within the age marking the present. Those ages are characterized by the basic zodiacal signs. Signs seen not as predictors of processes and outcomes, as one would see in a horoscope, but as symbols of the nature of the historical era it represents.

Does any of this sound familiar? For those with only the most cursory reading and understanding of the Book of Daniel, you will see the vision of Nebuchadnezzar in chapter 2. In a first person narrative, beginning in Daniel 2:5, Nebuchadnezzar tells of a vision he had of a giant human-like image —its head was gold, its chest and arms were silver, its belly and thighs of bronze, and its legs of iron and, feet of iron mixed with clay.

In Dr. Boulter's scheme, Virgo represents the transition point where the ascending phase transitions to the descending phase—but, represents the pinnacle of enlightenment. While marking the point of highest human accomplishment and planetary stability, it also marks the point where the long downhill slide starts. Conversely, Pisces represents the lowest point—the darkest times of human civilization and planetary stability, but also marks the start of the long slow ascent towards Virgo. According to Dr. Boulter, the Egyptians saw this entire cycle taking 29,000 years—fundamentally, a precessional cycle. Dr. Boulter's concept for Egyptian cosmology can also be seen in the 13 Baktun cycles of the Mayan long-count calendar and the 7 Yuga cycles of India.

So, where are we within this cycle at the present? According to Dr. Boulter, the last Golden Age was 11,500 years ago, corresponding geologically to the last Ice Age! We are 1500 years from the lowest point in the cycle. This might explain the societal and planetary upheavals that we see dominating the news these days!

Final note from the author -

As you will soon begin to understand, my view of history expands the typological concept of flexible time, mass extinction, and precession in a grand way. I will be dealing with history itself as a typological paradigm that is not fixed in a fatalistic inexorable linear process culminating in the apocalypse, but as a process that drives from apocalypse to rebirth and renewal, in a cyclical process that we may now be allowed to see in its fullness for the first time.

And, let me challenge you with this -

If God can create "the heavens and the earth" from nothing, by merely speaking them into existence, is it so hard to accept that He might be giving all of creation a second chance, maybe even a third, fourth, fifth or more chance? Romans 8:20-22 tells us,

> 20 *For the creation was subjected to futility, not willingly, but because of Him who subjected it in hope;* 21 *because the creation itself also will be delivered from the bondage of corruption into the glorious liberty of the children of God.* 22 *For we know that the whole creation groans and labors with birth pangs together until now.*

What does this mean? There is a very clear answer, and not the one you may be expecting. Prepare your feet to walk an old path in a new way.

PART 2: Genesis and Revelation

Psalm 19:1
The heavens declare the glory of God,
the sky proclaims His handiwork.

2.1 First Steps on the New Path

We all harbor our own "standard models," mental scaffolds built on the familiar and comfortable, shielding us from unsettling truths. The predictable offers solace, a sense that today will mirror yesterday, even if that continuity rests on shaky foundations, sometimes inaccurate, occasionally destructive.

Yet, what if the familiar obscures a grander reality? Let us dare to view old vexing enigmas through an unfamiliar lens, seeking answers to questions long dismissed.

Scattered across the earth are structures that defy our tidy histories: the pyramids of Giza; the Sphinx, gazing inscrutably through millennia; Stonehenge, aligned with solstices; the colossal stones of Baalbek's Heliopolis, weighing up to 1200 tons; and Göbekli Tepe's T-shaped pillars, etched with glyphs attributed to 9500 BC.

These megaliths, gigantic quarried stones, mock our assumptions about ancient capabilities, lacking clear evidence of their builders, methods, or dates. Who hoisted Baalbek's trilithon megaliths? How did Göbekli Tepe's builders, supposedly pre-agricultural, craft such sophistication?

A future book in this collection will probe these megaliths and possible Biblical ties, uncovering allusions deftly woven into the fabric of scripture—some highlighted, others subtly glossed over. The question of <u>who</u> built these wonders remains a tantalizing mystery, shrouded in the mists of possible lost civilizations. But the <u>when</u> they were constructed yields a clearer answer—they arose in a previous cycle—an epoch before our own, erased by time or cataclysm. Understanding cycles unlocks the enigma of their existence, placing them within a cosmic rhythm of creation and renewal.

The Bible, from the very first word in Genesis, to Revelation's closing vision of "a new heaven and earth" offers the key to discerning these cycles. Scripture is no mere chronicle but a celestial map, guiding us through history's cycles each marked by stars, empires, and divine purpose. By shedding the familiar, we may yet unravel the mysteries of our world's ancient stones and the timeless story they echo.

This book will not be an exhaustive treatise on prophecy or Biblical history. It is simply an effort to connect two ends of scripture and demonstrate the unity of the Biblical narrative. As I take you on this journey of exploration, we will view old artifacts in new ways.

This may be a bit confusing and at times unsettling —maybe even angering—but, stay with me to the end and I think you'll have a new perspective. Whether you accept it or not is your business.

You should find the contents of Table 2.1 (below) helpful.

Occasionally I will exchange views with someone whose opinion follows the course - ".. the Bible was written by men and they left a bunch of stuff out on purpose, so we don't have the whole story..." If this is similar to your opinion I hope you will discover the grand Architect of Time that I hope to reveal in these pages. That is, the God of Creation, who is certainly big enough to overcome human dispositions and make His will and desire known to us in spite of our shortcomings.

If you've ever said to yourself or anyone else for that matter, "If I were God I wouldn't do it that way..." When you try to make God in your own image your concept of God is woefully too small. God does not care much for how we would choose to do something. The God of Creation knows the "end from the beginning" (Ecclesiastes 3:11, Isaiah 46:10). There is a very specific plan and the Bible tells us what that plan is—but, you have to be willing to look deeply and invest the time and effort to understand what God's word actually says. It isn't always obvious, though it can be clear if we are willing to be still and prayerful enough to hear.

2.2 The Linguistic Origins

The main premise of this book is to demonstrate how the opening sequence of events in Genesis match and overlay with the closing sequence of events in Revelation. I noticed the correlations early in my study of Genesis, but did not think much of it at the time. It wasn't until I got well into a study of Daniel that I noticed something hidden in plain sight! That hidden gem was the possibility that the Bible was quietly concealing a pattern—a repetitive scheme that I hope to reveal to you as cycles.

When you read Genesis, especially the opening few chapters about creation and the first humans—did you ever ask yourself if Adam (and later, Eve) had a language? If so, what language was it and where did it originate?

We know Adam had a spoken language because, one of his first duties was to name all the animals (Genesis 2:18-20). Also, God walked and talked with Adam and Eve in the Garden (Genesis 3:8), and who really knows what kind of language they used, if they used language at all?

You will find some who are dedicated to the idea that the language of Adam and Eve, the very first language, would have been Hebrew. From a purely spiritual perspective that may be acceptable, but I doubt that we can ever really know what the pre-flood languages were like. With the flood, everything was reset except for the language and knowledge brought through the flood by Noah and his family. In due time, Noah's language would be scrambled at the Tower of Babel (Genesis 11).

In the process of history the world was repopulated; Abram (Abraham) was called out of Ur of the Chaldeans (Sumer), Isaac was born, then Jacob (Israel) was born. Subsequently, Jacob's son, Joseph, was born and sold into Egyptian slavery where God blessed and ultimately placed him in a position to rescue his family from a devastating famine.

The important question I would like to pose has less to do with the spiritual intrigue involved in this family saga, and much more to do with the practical matter of what languages were being spoken.

In the record to follow I will make reference to key events. These events will likely have contested dates for their occurrence. It is not the point of this book to dispute or defend any particular dating scheme. I offer them for reference only. As you will soon understand, I am much more concerned with the broad sequence of events and less the specific time in history in which they occurred.

James Ussher, Archbishop of Armagh (1625-1656), head of the Church of Ireland, set out to calculate the exact date of creation using only the names and ascribed ages of people and events listed in the Bible. Ussher's conclusion - creation happened around 6:00pm on October 22, 4004BC. This is the date that "young Earthers" use as their benchmark.

I have no particular reason to argue with Archbishop Ussher's chronology. However, by the time I finish revealing this theory in its fullness you will understand why I am not especially concerned about whether the earth is 'young' as Ussher proposes, or 'old' as science and archaeology suggest.

As a matter of reference, I present two Biblical chronologies below. Note that Archbishop Ussher's chronology is based purely on internal Biblical reference and reads as follows -

- 4004BC - Creation
- 2348BC - Flood of Noah
- 2242BC - Tower of Babel
- 1996BC - Birth of Abram.
- 1922BC - Abram leaves Ur (Sumeria)
- 1491BC - Israelites leave Egypt (Exodus)
- 1451BC - Moses' death, Israelites crossed into Canaan

- 1012BC - Construction of the First Temple
- 588BC - Fall of the Kingdom of Judah to Babylon

Compare this to more recent Biblical analysis that includes some factors that Ussher did not include. This timeline is provided by Dr. Douglas Petrovich, who uses timelines developed and refined by Edwin R. Thiele and Rodger Young. Petrovich's chronology reads as follows -

- 4174BC - Creation
- 2518BC - Flood of Noah
- 2234BC - Tower of Babel
- 2166BC - Birth of Abram
- 2091BC - Abram leaves Ur (Sumeria)
- 1446BC - Israelite leave Egypt (Exodus)
- 1406BC - Moses' death, Israelites crossed into Canaan
- 967BC - Construction of the First Temple
- 586BC - Jerusalem falls to the Babylons

I will revisit Dr. Petrovich in an upcoming section on the origins of the written Hebrew language.

Following the entry of Israel into the Promised Land (i.e. 1451BC or 1406BC, your choice), chronologies get a bit easier as family lineages and documentation become more detailed and accurate.

Having presented the two previous timelines, I would now like to recommend ignoring them altogether. It is my intention to put the Flood of Noah at the beginning of the last ice age—geologically, beginning approximately 12,500 years ago, ending approximately 11,000 years ago.

The global flood could explain the rapid growth of ice sheets on both poles. Make note of the fact that it snows very very little in Antarctica (South Pole) and Greenland, yet woolly mammoths were caught by surprise in a rapid freeze in the Arctic regions of Siberia and Canada. They were grazing with no particular since of urgency to seek shelter, but were frozen in place with food still in their mouths and undigested in their stomachs.

We need only acknowledge that the Ice Age happened and civilizations began to flourish afterward.

It is worth noting, Answers in Genesis (AiG), a well respected Biblical research foundation, places the Flood of Noah at approximately 2324BC (4350 years ago) and attributes the Ice Age to that flood event, as well.

AiG's mechanisms for the growth of the polar ice sheets following the global flood is well researched and very plausible. One of their leading experts on the subject, Michael Oard, has written extensively on this in his book "Frozen in Time", showing how the post-flood ice age explains woolly mammoths and other ice age mysteries within a biblical framework.

Emergence of the First Languages

The purpose of establishing a rough timeline is to give context to the emergence of languages after Noah and his family landed in the mountains of Ararat. This is a topic I have never heard taught in churches. Though the Tower of Babel story is familiar and common there are some more broad and significant questions about how the Hebrew language, spoken and written, developed after the Tower of Babel events.

Biblical timelines can be somewhat difficult to follow. We tend to see consecutive stories occurring with no particular way to place them in a formal historical sequence.

As demonstrated in the timelines shown previously, Abram's family was called out of Ur some 200 to 300 years after the Tower of Babel event. This certainly gives ample time for new cultures with emerging languages to develop and provides the context for the divergence of languages around the world.

Later in the Biblical narrative (Genesis 17:5), after receiving the promise of God, Abram is renamed Abraham. I will refer to him as Abraham going forward.

An interesting coincidence is the sudden and inexplicable presence of the Sumerians and the Sumerian language in Mesopotamia. Mesopotamia is the "land between the rivers"—those rivers being what we know as the Tigris and Euphrates Rivers just north of the Persian Gulf in an area we know today as Iran. It represents the eastern limit of what is considered The Fertile Crescent and the Cradle of Civilization and is a land of incredible antiquity and the beginning of many remarkable innovations that are present with us today.

The ancient Sumerians of Mesopotamia gave the world its foundational building blocks of civilization. They invented the first writing system (Cuneiform), developed the earliest known legal codes, and pioneered urban city-states. Their base-60 mathematical system still governs our 60-minute hour and 360-degree circle. They established the first organized religion, agriculture-based economies, large-scale irrigation systems, the wheel, and the 12-constellation zodiac — a celestial framework to be explored more fully later in this book. They developed the first monumental architecture like ziggurats, the architectural pattern for the Tower of Babel which was built earlier in this same area.

For all the breakthrough innovations established in ancient Sumer, there was a very dark and insidious side to that early society. It is the side that God was calling Abraham and his family to separate themselves from and to establish a people unto Himself.

Dr. Heather Lynn, in her book "Evil Archaeology", dedicates the entirety of chapter 2, "The Cradle of Evil" to humanity's post-flood return to the evil and God-less behavior that provoked the flood to begin with.

Following the Tower of Babel, languages were scrambled and nations would begin to form around people with a common language and the ability to understand one another. Over time, as the Sumerian language began to wane in common usage, it was replaced by a proto-Aramaic dialect ("proto" meaning, first or original). Abraham and his family likely would have spoken some form of proto-Aramaic, as the original Sumerian language itself became the language of administration, literature, religion, and royal inscriptions in Ur, while its use as a daily spoken language was in decline.

I think it's safe to say, Abraham and his family spoke some form of Aramaic. So then, where did Hebrew come from and how did Abraham's family come to speak a different language than their progenitor?

Well, I'm glad you asked!

2.3 The Foundations of Written Hebrew and the Bible

If you didn't do well in primary school English class, this might bore you to tears! I really didn't like English until I took a Classical Greek course in college. It was only then that I discovered the need and developed an appreciation for all the complexity and nuance of English grammar. Recently, I have also grown an affinity for linguistics, the study of languages and their origins.

The point of this section is to understand how the Hebrew language originated, how Moses had access to a developed phonetic alphabet and could easily have written the Torah in 40 years. This is the plan and purpose that defines God's magnificent patterns in history and nature, and reveals the cycles that are the critical point of this book.

In the following sections I will lean heavily on the work of Dr. Douglas Petrovich, an accomplished Egyptologist, linguist and expert in Near Eastern history. Dr. Petrovich has studied and researched extensively on the origins of the Hebrew language and alphabet. Though I find Dr. Petrovich's work very compelling, he is by no means the only valuable resource for the study of Hebrew history and timelines or the development of the Hebrew language.

Among other valuable sources are:

Dr. David Rohl, Egyptologist and Near Eastern historian proposes a radical reexamination of traditional Near Eastern chronologies. Dr. Rohl's work seeks to better contextualize Biblical timelines and generally accepted ancient chronologies. His work as an Egyptologist has helped to identify many sources of proto- or paleo-Hebrew, the original and developmental forms of written Hebrew, as they appear in the context of Egyptian archaeology.

Timothy Mahoney, investigative filmmaker, author and producer of many in-depth investigations have sought to resolve many old questions about the veracity of the Bible and traditional Biblical chronologies. Tim's work synthesizes the efforts of Dr. Petrovich, Dr. Rohl and numerous others to present a new and coherent story of Biblical history and the origin and development of written Hebrew.

For simplicity and clarity, I will refer to the original forms of proto- or paleo-Hebrew as "Old-Hebrew" going forward in this book. Keep in mind, Hebrew is written and read right-to-left.

Further Steps on the New Path

To understand the development of cultures, nations, and languages within a Biblical framework we have to keep in mind a very basic sequence of events -

1. The corrupt world was reset by a global flood
2. Noah, his three sons—Shem, Ham and Japheth —and their wives repopulate the world
3. Cultures and languages are reset at the Tower of Babel

It might be helpful to read Genesis 11 before Genesis 10, as chapter 11 tells the story of the Tower of Babel and the making of nations and languages, while chapter 10 reveals the effect on each lineage of Noah's sons (10:5, 20 & 31), "..according to their clans and languages, by their lands and nations." In chapter 10 the results of chapter 11 are already present.

After the waters of the global flood receded Noah's family was able to establish themselves as a stable community. In due course, there must have been ample time for the families to migrate eastward and southward from the mountains of Ararat.

We are told in Genesis 11:2 that the unified peoples of the world journeyed <u>from</u> the east to the plains (or valley) of Shinar (Sumer) and dwelt there.

Grammatically, the case can be made that the people migrated to the eastern portion of Shinar, presumably from the west, and settle in that land.

At this point in history they are all likely speaking some variant of what would become Sumerian. I suggest that this variant was spoken in the pre-flood world and carried over into the early post-flood world.

At this time Shem, Ham, Japheth and their children can all talk to grandpa Noah without any trouble or miscommunication, but grandpa Noah, though genetically untainted from the pre-flood world, is not able to keep his family on the straight-and-narrow. People begin to revert to their base impulses very quickly. While the flood reset purified the bloodline, it didn't purify the rebellious nature of people and they began to separate themselves from God's presence almost immediately. In a short time they are building a ziggurat (tower) to reach God, not as an act of devotion, but an act of defiance. The Tower was intended to dethrone God and bring Him to earth as they elevated themselves to His position. They got what was coming to them—confusion and alienation—from God and each other!

Genesis 10 tells the story of family diffusion after the Tower of Babel event. As illustrated in Map 2.1,

Japheth's family line migrated primarily to the north and eastward into Asia and India, but also westward into what would become Europe.

Japheth and the Proto-Indo-European Horizon

If Noah's ark came to rest in the highland region between the Black Sea and the Caspian Sea — in the general vicinity of what is today eastern Anatolia (Turkey) and the southern Caucasus — then the dispersion of Japheth's descendants northward and westward traces a path that corresponds remarkably well with the proposed spread of what modern linguists call the Proto-Indo-European (PIE) language root.

The traditional Hittite homelands of central Anatolia represent one of the earliest attested branches of the PIE language family tree. From there, the language and its daughter tongues radiated outward: westward into Greece, Italy, and eventually all of Europe; northward into the Pontic-Caspian steppe, where the Kurgan hypothesis locates a secondary explosion of PIE speakers; and eastward into Persia, the Indian subcontinent, and Central Asia.

Genesis lists among Japheth's descendants peoples identifiable with the Medes, Greeks, Thracians, and coastal Anatolian groups —

precisely the populations whose languages later emerge as Indo-European. The geographic coherence between a northern Ararat landing site and this dispersal pattern is one of the more compelling geographic alignments available to a Biblical framework of language origins.

Ham and the Languages of Africa

Ham's family migrated further into Africa and are notable as the Egyptians of Biblical history. Linguistically, Ham's name [H2526 or H2525] means - warm, hot; suggesting the climate of Africa. Psalm 106 recalls "..(God's) wondrous deeds in the land of Ham, awesome deeds at the Sea of Reeds.." recognizing God's deliverance of the Hebrew people from captivity in Egypt.

In their book, "Black Genesis: The Prehistoric Origins of Ancient Egypt", Robert Bauval and Dr. Thomas Brophy layout the foundation for development of Biblical era Egypt by the black peoples of Hamitic Africa. However, as I develop my theory of biblical cycles in this book, it should become apparent that the great pyramids and necropolis complexes, especially those of Saqqara, Giza, Dahshur and similarly dated architecture, were not built in the biblical time period and the Hebrew/Israelite people had nothing to do with pyramid construction. The pyramids had been present in Egypt many millennia prior to the Israelite sojourn in Egypt.

As I have suggested, the pyramids predate the era of the biblical narrative. I also imply that hieroglyphics predate the same, as hieroglyphics as a written language predate the pyramids. This is critically important to the understanding of the origins of the written Hebrew language—the written Hebrew that Moses would use to write the Pentateuch, the first five books of the Hebrew Bible.

This is where the language framework encounters its most significant challenge, and intellectual honesty requires stating it plainly: there is no established proto-African language. Africa is home to four major language superfamilies—Afroasiatic, Niger-Congo, Nilo-Saharan, and Khoisan—and no credible linguistic reconstruction has unified them under a single ancestral tongue.

Unlike PIE, which has been reconstructed with considerable rigor, or Proto-Semitic, which is well-attested, a "proto-African" proto-language does not exist in the scholarly literature and likely never will, given the enormous genetic and typological diversity among these families.

What *can* be said is this: the descendants of Ham in Genesis 10 — Cush, Mizraim, Put, and Canaan — correspond broadly to the regions of sub-Saharan Africa (Cush), Egypt (Mizraim), and the Libyan or Somali coast (Put). Importantly, the Afroasiatic family, which includes ancient Egyptian, Cushitic languages (spoken in the Horn of Africa and Sudan), Berber, Chadic, and Semitic, does provide a partial linguistic bridge. Egypt and Cush — two of Ham's named sons — occupy exactly the African territories where Afroasiatic languages are indigenous. This is a genuine and noteworthy correspondence, though it simultaneously complicates the Shem/Ham boundary, since Semitic languages are also Afroasiatic and are typically attributed to Shem's line. This is where an important overlap takes place.

Shem and the Semitic Languages

Shem's family stayed closest to the homeland and occupied the areas of Shinar and what we would come to know as the Arabian peninsula and the Levant (Turkey, Syria, Lebanon, Jordan and Israel). In the course of time, the Shemites would migrate westward, as Abraham's family was called to separate itself from the evils of Ur. Further, Abraham's lineage—Isaac and Jacob (who was later renamed Israel) had sons who became the twelve tribes of Israel. And, as family dynamics tend to happen in the real world, petty jealousy consumed the older ten sons of Israel and they sold the eleventh son, Joseph, into slavery in Egypt, hoping never to see him again. Subsequently, a twelfth son was born to Jacob, Benjamin.

To abbreviate a long inspiring story of redemption, Joseph comes into significant power and influence as vizier in the pharaonic kingdom of Senusret II (also called, Sesostris II); according to Dr. Petrovich's timeline, 1880-1875BC (Egypt's 12th Dynasty, Middle Kingdom). From Joseph's position of influence and privilege in the house of the pharaoh, along with one of his two sons, Manasseh, Joseph begins to derive and codify a system of phonetic substitutes for common hieroglyphic symbols. See Chart 2.1 for the basic progression of the substitutions as they develop into Old-Hebrew. The left-most column of the chart shows the block, or square, Hebrew letters that developed in Babylon during the time of Daniel and are most familiar to us today. More on that to come.

Joseph's family was to be honored and accepted by the pharaoh and granted some the most prime land in the pharaoh's kingdom, the Land of Goshen in the lower Nile delta. After about 300 years in Egypt, there arose a pharaoh (Exodus 1:8), "..who did not

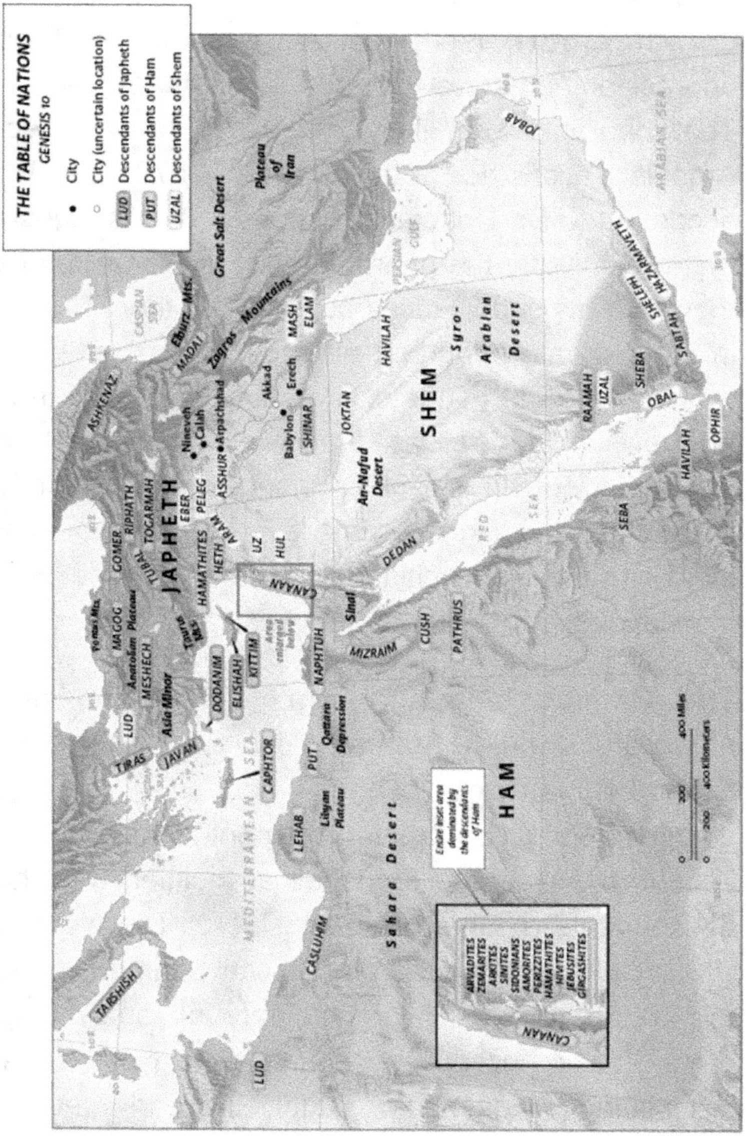

Map 2.1 - Table of Nations
Genesis 10, post-flood broad view.
Holman Bible Publishers ©2007

know Joseph..", was threatened by the prodigious growth of the Israelites and called for their oppression and slavery for about another 100 years. That pharaoh was likely Ahmose I (also called, Amosis I) of the 18th Dynasty 1570-1546BC, according to Dr. Petrovich. Exodus 1:8-14 suggests that not only did the pharaoh afflict the Israelites as they built his supply cities but also put them to task in other areas. Important among those other tasks was working the mines of the southwest Sinai. It is in these mining areas that we find compelling inscriptions of hieroglyphics alongside Old-Hebrew lines of text bearing the same level of wear on each, indicating they were chiseled at the same time. See Inscription 2.1.

The Semitic language family—encompassing Akkadian, Ugaritic, Aramaic, Hebrew, Arabic, and Ethiopic, among others—has its earliest documented presence in Mesopotamia and the Levant, precisely the territories associated with the descendants of Shem in Genesis 10. The work of Dr. Petrovich has added a provocative dimension to this picture. Petrovich argues, through analysis of proto-Sinaitic inscriptions found in Egypt's Sinai Peninsula and dating to roughly the early second millennium BC, that the world's first alphabetic script was composed in a language identifiable as early Hebrew, likely by Semitic-speaking laborers present in Egypt during a period corresponding to the Israelite sojourning.

If his identification is correct, the Old-Hebrew alphabet—and by extension the ancestor of nearly every alphabet in use today—originated not in Canaan but in Egypt, carried there by Semitic migrants and crystallized into writing within an Egyptian context. This would place the birth of alphabetic writing squarely within a Shemitic population, consistent with the Biblical genealogy, while explaining its Egyptian derivation through historical migration rather than indigenous development.

Moses and the Torah

Inscription 2.1 demonstrates an example of hieroglyphics alongside Old-Hebrew, but this is not an isolated sample. Similar examples can be found elsewhere in the Sinai mining regions, along the course of the Nile river, and in the Nile delta's ancient Goshen region.

With his privileged position and education within the pharaoh's inner circle Joseph had ample opportunity to identify a need and develop a more accessible written language than that of hieroglyphics. Hieroglyphics, as a written form with its many variants and nuances, would make it impractical for long narrative such as the Torah. Instead, Joseph developed a highly versatile and easily learned phonetic script that became the basis for most written and spoken languages of the Levant. Contrary to popular belief, the Phoenicians did not develop the first phonetic alphabet. The Hebrews did, and through long and diverse interactions shared it with most of the rest of the Near Eastern societies.

The Israelites spent over 400 years in Egypt, some of that time as privileged guests and some as oppressed. This would be ample time to develop the alphabet that Moses would use, and there is little doubt that a fully developed phonetic alphabet existed for Moses to pen the Torah. With 40 years of wandering, most of which were stationary camping periods—the Israelites did not trek the desert continuously for 40 years—Moses had plenty of time to compose the entirety of the first five books of the Bible—the Torah; in Greek, the Pentateuch.

The story of Moses, his rescue and ascension to authority among his Hebrew brethren are recounted in his own words in the book of Exodus. Moses, like Joseph, was a genetic Hebrew growing up in the safety and entitlement of pharaoh's house. Having inherited a nearly 400 year legacy of Joseph's breakthrough phonetic writing technique, writing would have been common among royal elites of this time.

As we will learn in many of the historical books of the Tanahk (found in the Nevi'im, the historical and prophetic books of the Hebrew Bible), the basics of spoken Hebrew-like languages—known collectively as proto-Canaanite, or proto-Siniatic—were familiar throughout the Levant by the time of Israel's establishment in the Promised Land. Though individual family groups and tribes may have spoken their own dialectic version of proto-Canaanite they could understand the fundamentals as they communicated within the Levant. In this way we understand how the Israelites came to speak a different language than Abraham as he left Ur centuries prior.

Notable examples of Semitic language overlap are as follows:

1. Rahab and the Israelite Spies — Joshua 2:1–21

When Joshua sends two spies into Jericho, they lodge with Rahab, a Canaanite woman, and engage in an extended, substantive conversation with no mention of any interpreter. Rahab demonstrates not only linguistic fluency with the Israelites but a sophisticated theological awareness, saying "the LORD your God, He is God in heaven above and on

Chart 2.1 - Consonantal Hebrew Alphabet
Progression from semi-Hieroglyphic to Old-Hebrew
("Sinai ###" refer to inscriptions with that symbol)
Dr. Douglas Petrovich ©2016

earth beneath" (Joshua 2:11). This is perhaps the most direct early evidence of seamless communication between Israelites and indigenous Canaanites.

2. The Gibeonite Negotiation — Joshua 9:3–15

The Hivites of Gibeon devise a deception requiring them to engage Joshua and the leaders of Israel in detailed, persuasive diplomatic negotiation, entirely without an interpreter. They construct a plausible narrative, answer follow-up questions, and successfully conclude a binding covenant treaty. The sophistication of their communication — including culturally resonant references to Israel's God and their military reputation — strongly implies a shared or mutually intelligible linguistic framework between Hivites and Israelites.

3. Jephthah's Diplomatic Exchange with Ammon — Judges 11:12–28

This passage is particularly compelling as, Jephthah, the Gileadite judge, conducts what amounts to a formal written diplomatic correspondence directly with the king of Ammon, arguing historical land claims across multiple exchanges. The dialogue is historically detailed and legally structured, with each side fully comprehending the other's arguments. That such

Inscription 2.1 - from the mines of Sinai
(This inscription is to be read from the long side of this page.)
Sinai 46 (left) hieroglyphics
Sinai 377 (right) Old-Hebrew

nuanced legal and historical reasoning could be exchanged without any noted linguistic barrier speaks strongly to a shared Semitic linguistic heritage across the Transjordanian region.

4. Ruth the Moabitess — Ruth 1–4

Though sometimes categorized separately, Ruth sits within the historical narrative of the Judges period and offers perhaps the most intimate portrait of cross-cultural linguistic fluency in the entire Hebrew Bible. Ruth, a Moabitess, communicates seamlessly with Naomi, Boaz, the town elders, and the women of Bethlehem at every level — emotionally, legally, and socially — with no hint of any language barrier whatsoever. Given that Moab sat east of the Dead Sea and had its own distinct national identity, this frictionless communication is highly suggestive of the broad reach of a common Hebrew/Canaanite linguistic root.

5. David and Achish, King of Gath — 1 Samuel 21:10–15 & 27:1–12

David's two extended stays among the Philistines of Gath, communicating directly and repeatedly with King Achish, are notable because the Philistines are typically considered a Sea People of Aegean origin — and yet by the time of David, Achish and his court communicate with David with complete fluency and apparent ease. David even serves in Achish's military and receives the town of Ziklag as a grant, all negotiated in direct conversation. This suggests that whatever the Philistines' original Aegean linguistic roots may have been, by the early monarchic period in Israel they had thoroughly adopted the Hebrew/Canaanite-based tongue.

Other such examples of cross-cultural discussions and negotiations abound with only the most cursory reading of the historical books of the Hebrew Bible.

The salient question we now have to answer is, grammatically how well developed and nuanced was Old-Hebrew at the time Moses began to pen the Torah? While there is no definitive answer to this question, the likely conclusion is, Moses probably derived the final version of Biblical Old-Hebrew while ensconced in his tent in the deserts of the Sinai.

The following points show how we got from the version that Moses used to the version that is used today:

- Old-Hebrew (proto-Hebrew or, paleo-Hebrew) - this is the alphabet Moses would have used to write the Torah, and the alphabetic script of the tablets of the Ten Commandments as they came down from Mt. Sinai.

    - Example: 𐤁𐤓𐤀𐤔𐤉𐤕 - Old-Hebrew had no vowels and would have originated in Egypt during the time of, and most likely by, Joseph as vizier of Egypt.

- Post-exilic block, or square, Hebrew / Aramaic (Babylonian) forms - this is the highly modified version of the Mosaic Old-Hebrew alphabet that developed in Babylon during the time of Daniel. This version had no vowels and would have originated sometime after 586BC, the beginning of the Exile of Judah. This is the written Hebrew form most like that used in the public square in Israel today.
    - Example: בראשית - block Hebrew before the addition of diacritical vowel markings.

- Diacritical block Hebrew - this is the Hebrew/ Aramaic block form alphabet that had vowel pointing assignments within the letters and words. These important additions to written Hebrew developed primarily in Tiberias, on the western shore of the Sea of Galilee c. 500-950AD by the Masoretes. This is the alphabetic form we see in religious Hebrew texts today.
    - Example: בְּרֵאשִׁית - block form Hebrew with diacritical vowel markings. It is worth

noting that the diacritical vowel markings were carefully placed in and around the original block Hebrew text in such a way that they did not interfere with or obscure any of the original text.

The first word of Genesis, penned by Moses, would have looked much like this in Old-Hebrew - 𐤁𐤓𐤀𐤔𐤉𐤕 . The Old-Hebrew and the block form script shown in the previous paragraphs (בראשית) are the same word - b'ra-sheet, the very first word in Genesis 1:1. It is translated to mean, "in (the) beginning". That single Hebrew word requires up to three words in English, but there is some uncertainty as to what those words are exactly!

Because understanding the very first word in the Hebrew Bible is essential to understanding the interpretation of Biblical cycles I will spend some extra time on it. It is not my intention to get into Hebrew grammatical weeds, but it is important to understand why the English Old Testament is translated differently than the English version of the Hebrew Biblical text (JPS).

The most fundamental reason our English Bible translates the very first word in the Hebrew Bible as "In **the** beginning.." (emphasis added) is because the Septuagint, the original Greek translation of the

Hebrew text, translates it that way. Our English Old Testament has traditionally taken many of its meanings and basic understandings of ancient Hebrew from the Septuagint version and, in this case, we tend to default to that translation of the text. However, is that the best approach to the Hebrew text? As you will soon understand, some ancient and contemporary Hebrew textual scholars might disagree.

The key to unravelling the true mystery of creation, and why the story ends in Revelation as it does, has everything in the world to do with the very first word in the Hebrew Bible, the first word in the Torah. The word we translate as "In the beginning", is a single Hebrew word—in English pronounced, b-ray-sheet.

In Hebrew the word "b'ra-sheet" [H7225] does not have the definite article 'the' included or implied in its grammatical structure and is in the grammatical *construct form*. This construct form links it to the activities that follow, which of course makes a lot of sense, but it also implies that heaven and earth were already present as the process of creation got started and had already become "formless and void". See Section 2.4 Without Form and Void, for further analysis of this very important word combination.

For a moment, setting aside our traditional Christian translation of Genesis 1:1, here are some other approaches to that opening line:

- The Hebrew Bible (JPS) translates Genesis 1:1 as "When God began to create heaven and earth.."

- In his translation <u>The Schocken Bible: The Five Books of Moses</u> (1995) Everett Fox says, "At the beginning of God's creating of the heavens and the earth.."

- In his translation <u>Genesis: The JPS Torah Commentary</u> (1989) Nahum Sarna says, "At the beginning of (the process of) creation.."

- Dr. Douglas Petrovich, ancient Hebrew expert, also questions the presence of the definite article. See his Patterns of Evidence: Hebrew the World's Oldest Alphabet video series from. Thinking Man Films.

- A less nuanced reading directly into English as Moses might have actually meant in Old-Hebrew, "In **A** beginning…" The indefinite article is implied by the lack of a definite article "the". Hebrew has a definite article—similar to the word "the" in English—and its absence implies an indefinite article "A" as used in some western languages. That definite article does not appear in the word “b'ra-sheet” and is not implied in its construction.

- Finally, Dr. Josh Wilson, writing for Answers Research Journal, a publication division of Answers in Genesis (AiG); Answers Research Journal (2018):1-21 Linguistic Traits of Hebrew Relator Nouns and Their Implications for Translating Genesis 1:1, concludes . . he’s not really sure about the definite/indefinite articles either. Writing extensively on the topic of construction of Genesis 1:1, specifically on the word in question—b’ra-sheet [H7225]—Dr. Wilson, Professor of Bible at Missouri Baptist University, spends 21 pages of heavily footnoted, referenced and bibliographied study as he critically analyzes all aspects of this important verse and word. Not completely satisfied with his uncertain conclusion he revisits the topic in Answers Research Journal (2018):341-358 Syntactical Features of Hebrew Genitive Clauses and Their Implications for Translating Genesis 1:1. This time, he’s more satisfied with the conclusion that the definite article “the” is more acceptable.

What is clear about this unique Hebrew word is that it is very flexible. But, most importantly, it is the first clue we have that implies repeating cycles and/or a repetitive paradigm in the hidden history of planet earth and probably the entirety of creation.

While definitive proof is difficult to come by, of what Moses could have written and meant, as he wrote in significantly less grammatically and syntactically developed Old-Hebrew, I doubt Moses exerted any particular effort in his word choice. I submit now - **Moses very likely recognized the repetition in history and did not need a definite article to make the narrative fit into linear time!** Moses was taking dictation from the intensions of an infinite God who is quite capable of restarting the process of creation as often as He cares to.

The story of Genesis 1 is creation beginning, but not for the first time. The reason becomes clear in the narrative of the entire Book of Revelation. If we imagine those apocalyptic events as being the cause of Genesis 1:2, "..the earth being unformed and void, with darkness over the face of the deep..", we have the connecting narrative and better understand why the creation story starts as it does, and why Revelation ends as it does. Revelation 21:1 says, "Now I saw a new heaven and new earth, for the first heaven and the first earth passed away." The new one replaces the old, as described in Genesis 1. Therefore, when it ends in Revelation it begins anew in Genesis . . so, now we understand **Revelation to Genesis** is the patterned cycle!

2.4 Unformed and Void

Genesis 1:2 (JPS), "..and the earth being unformed and void.."

In Hebrew, "..וְהָאָרֶץ (v'h'erets) [H776] הָיְתָה (hayatah) [H1961] תֹהוּ (tohu) [H8414] וָבֹהוּ (v'bohu) [H922].."

Many diligent theological seekers have sought to resolve a nagging question offered in this phrase of the Genesis creation story and suggest that something serious, maybe even cataclysmic, must have happened just before the event described as "..a first day" (Genesis 1:5, JPS). Note, in the Hebrew Bible it is not "..*the* first day". The most important question offered here - why would an omnipotent God, capable of creating something from nothing, create a world that required further repair?!

Prominent among the theories offered to answer this question is the Gap Theory. The Gap Theory suggests there must have been something else going on in the heavens and the earth at the time of creation. Chief among those suggestions was the rebellion of Lucifer, his minions and their fall, and that rebellion is what caused things to be

"..unformed and void, with darkness over the surface of the deep..."

I would offer that *something else* was, more likely, the events described in Revelation—what we call the Apocalypse. Revelation offers the complete list and sequence of destruction that Genesis describes as re-creation and repair. Therefore, **Revelation is the complete account of creation becoming "unformed and void"**.

Here again, is a grammatical challenge. The question is not the meaning of "..unformed and void" —in Hebrew: tohu [H8414] v'bohu [H922]. The 'v' prefixed to 'bohu' is the Hebrew article for "and". The question is—when did it happen? In other words, what is the meaning of the Hebrew word הָיְתָה (hayatah) [H1961]? Did the chaos happen after creation began or was it already present. Grammatically, it can go either way and there is some level of debate as to which it should be. You can imagine where I'm going with this . . it was already in place. In the non-linear cyclical paradigm, creation "was", from the end of the previous cycle, already in a state of apocalyptic chaos, needing restoration and repair!

Further Down the Path

At first glance, the creation account in Genesis 1 perplexes the modern mind. God commands, "Let the earth sprout vegetation: seed-bearing plants, fruit trees of every kind on earth that bear fruit with the seed in it" (Genesis 1:11), crafting verdant life on the third day, yet the sun—the celestial engines of photosynthesis—and the moon emerge only on the fourth day of creation. Stranger still, God proclaims, "Let there be light [H216]" (Genesis 1:3), bathing the void in radiance—but the sun, our presumed source of light, awaits its debut four days hence. This sequence defies our earthly logic, a cosmic puzzle that seems to invert the natural order.

Yet, this anomaly is no oversight; it is a deliberate clue, etched into scripture's opening verses. Turning to Revelation chapters 21 & 22, the pattern reappears, as if mirrored across eons of time. In Revelation 21:23, "the city had no need of the sun or of the moon to shine in it, for the glory of God illuminated it. The Lamb is its light." Here, divine radiance precedes celestial bodies, just as in Genesis. Revelation 22:2 adds, "the tree of life... yielded her fruit every month," echoing Genesis' pre-solar vegetation, thriving under God's eternal provision. The parallels are striking - light before sun, life before luminaries—suggesting not a haphazard narrative but a profound connection.

This connection is more than intriguing; it is foundational to grasping the rhythmic cycles that pulse through Earth's history. This book posits that Genesis 1 through 3 and Revelation 18 through 22 are not isolated bookends, but markers of a repeating saga—a cosmic dance of creation, fall, judgment, and renewal—played out on our familiar blue planet. Each cycle begins with divine light—John 1:4, "..the life was the light [G5457] of men."—that sustains life through God's glory, not mere physics.

More importantly, the Hebrew word בְּרֵאשִׁית (b'ra-sheet [H7225], Genesis 1:1, suggests something even more profound about the creation process. Our English understanding of this word betrays its true meaning. In the Hebrew Bible (JPS) it reads, "When God began to create", hints at "a beginning," one among many. While in Revelation 21:1 we see "new heaven and earth" as it replaces a "first" that passed away. More will be revealed on these ideas later.

These scriptural echoes reveal earth as a stage for a recurring drama, where God's light precedes creation's mechanics, and life flourishes by His will.

The stars, ordained as "signs" [H226] (Genesis 1:14), chart the echo of cycles, from Virgo's dawn through to Libra's scales of justice. Far from a linear tale, Genesis and Revelation unveil a repeating cycle, where each turn renews the eternal story, God's glory illuminating our world, cycle after cycle, until the final renewal dawns. Have we arrived at that final cycle? I'm sure all previous generations have asked that eternal question.

What then of Revelation? Revelation chapters 19 through 22 speak of the Bride of the Lamb, and the New Jerusalem being the bride (Revelation 19:7-9, 21:2, 21:9-11, 22:17; also Isaiah 62:5, Matthew 9:15, John 3:29, 2 Corinthians 11:2 & Ephesians 5:25-33). In these contexts the bride is typically interpreted as the church or the assembly of believers, as they are betrothed to the Lamb of God. As the body of believers wed to the Lamb, our Savior Jesus Christ, we are once again allowed eternal life (1 John 3:2 & Revelation 21:1-4). As in the Garden of Eden, we will be in the presence of God with His abundance and provision around us—the river of life flowing from His throne, watering the tree of life growing tall and wide, and the nations shall be healed and the people of faith will live forever (Revelation 22:1-5, Genesis 2:8-10).

Table 2.1 reveals more connections and correlations between the creation story of Genesis and resetting the stage for creation in Revelation.

Table 2.1: Comparative Events Genesis & Revelation

GENESIS Tanakh (JPS)	REVELATION (NKJV)	Correlation
1:1 - "When God began to create heaven and earth.."	21:1 - "I saw a new heaven and a new earth."	A fresh creation cycle begins.
1:2 - "..and the earth (became) was formless and void.."	Ch. 18 - 20 The chronicle of events that destroyed the original state of creation.	Revelation is the story of how "formless & void" happened. **(See sec. 2.4)**
1:2 - "..and darkness was over the surface of the deep.."	8:12 - "..and a third of..sun..moon..stars.. were darkened.."	Revelation is the story of how the "darkness" happened. **(See sec. 2.4)**
1:3-4 - "Let there be light"; and there was light.	21:23 - "The glory of God illuminated it"	Light precedes celestial bodies, the Lamb is the Light.
1:6-8 - "..let there be an expanse to separate the waters.."	21:1 - "Also there was no more sea."	The firmament and waters represent chaos and disorder.
1:11-12 - "Let the earth sprout vegetation.."	22:2 - "The tree of life . . yielding her fruit"	Vegetation thrives by divine Light, precedes the sun.
1:14-16 - "Lights in the expanse . . serve as signs . . the two great lights.."	21:23 & 22:5 - "No need for the sun and the moon .. The Lamb is its Light."	Light shifts from objects of creation to heralds of God's glory.

GENESIS Tanakh (JPS)	REVELATION (NKJV)	Correlation
1:20-25 - "..swarms of living creatures . . be fertile and increase.."	21:5 - "Behold, I make all things new."	All things are encouraged to be fruitful and multiply.
1:26-27 - "..let Us make man in Our image.."	21:3 - "..the tabernacle of God is with men.."	Adam is like the angels, created by God for God. **(See sec. 2.5)**
1:31, 2:17 - "God saw all that He made, and found it very good"	21:4 - "No more death, nor sorrow"	Perfection restored, the cycle has restarted.
2:1-3 - ".. God finished the work . . and He ceased . . from all the work He had done."	22:3-5 - "..the throne of God and the Lamb shall be in it."	The throne represents the final place of rest, Jesus sat after His work of salvation was completed on the cross.
2:9 - "..God caused to grow every tree . . with the tree of life.."	22:2a - "..on either side of the river, was the tree of life."	Restoration and peace, new and abundant life.
Gen. 10 & 11 - The nations are divided and languages confused.	22:2b - "The leaves were for the healing of the nations."	Healing of the nations, renewed unity among humanity, one language restored.
2:10 - "..a river went out of Eden to water the garden."	22:1 - "..a pure river of life . . from the throne of God."	The river represents abundance and life, flowing from God.

GENESIS Tanakh (JPS)	REVELATION (NKJV)	Correlation
2:22-24 - "The Lord God fashioned the rib . . into a woman.."	21:2 "..prepared as a bride adorned for her husband."	This represents the unity of mankind together with Him, all completely devoted to Him.
2:16ff - God walked with man in the garden, implied throughout the text	21:3 - "the tabernacle of God is with men . . He will dwell with them."	God's desire has always been to walk freely with us in His creation. **(See sec. 2.5)**
2:21-24 - "This is bone of my bones . . a man shall leave . . . and clings to his wife.."	19:7-9 - ".. for the marriage of the Lamb has come .. blessed are those who are called.."	We are no longer in need of marriage as we will be married to our Savior. **(See sec. 2.5)**
2:8-9 & 3:1 - "..and the tree of life.." & "Now the serpent (Satan) was the shrewdest.."	20:1-3 & 7-10 - Satan is bound but then released to influence God's creation.	Satan was present in the Garden from the beginning of the creation story. **(See sec. 2.6)**
Genesis 10 - languages confused leading to the division of nations.	22:2b - "..The leaves of the tree *were* for the healing of the nations."	In the new Jerusalem we will have one language again.
Matt. 24:34 - "..this generation will by no means pass away till all these things take place."	Rev. 22:7 & 20 - "..I am coming quickly. Blessed is he who keeps the words of prophecy.."	"This generation" is the cycle we are currently in, as humanity awaited His return.

GENESIS Tanakh (JPS)	REVELATION (NKJV)	Correlation
Matt. 24:35 "Heaven & earth will pass away, but My words will by no means pass away."	Rev. & Gen. - The new heavens and the new earth created by the spoken Word of God.	Jesus said this would happen. But, He is about restoration by His Word of life.

2.5 Who was Adam (and Eve)

Critically important in the cyclical process, but not necessarily obvious, we turn to observations about Genesis and Revelation and look at marriage and the concept of sexual reproduction. There are some profoundly interesting things Jesus revealed in the Gospels that bear directly on the question of what human beings will become as the cycle turns from Revelation back to Genesis.

When confronted mockingly by Sadducees in Matthew 22:23–33 (also Mark 12:18–27 and Luke 20:27–40) with a story about resurrection—the Sadducees did not believe in resurrection—Jesus was asked about the marital status of a woman after death. His response was both simple and staggering: "..in the resurrection they neither marry nor are given in marriage, but are like the angels of God in heaven." (Matthew 22:30)

What is being said here is provocative and complex. What does it mean to be like the angels? And what does that tell us about marriage, reproduction, and the ultimate destiny of the human race?

Angels: Created, Genderless, and Ever in God's Presence

To understand where humanity is headed, we need to understand what angels are. The Old and New Testaments are full of references to the activities of angels—they deliver messages, carry out judgments, worship before the throne of God—yet they tell us surprisingly little about their fundamental nature. What we do know is this: angels are created beings who do not reproduce. God makes them as He wills, and they exist in a mode of being entirely different from fallen humanity.

The primary Hebrew word for angel is מַלְאָךְ (malak) [H4397], meaning messenger or representative. The Greek equivalent is ἄγγελος (angelos) [G32], carrying the same root meaning. In neither language does the term carry any inherent gender. The vast majority of angelic beings described in Scripture appear in masculine grammatical forms because biblical Hebrew and Greek default to the masculine in ambiguous cases—not because these beings are biologically male. Where angels appear with physical characteristics in visions, such as the seraphim of Isaiah 6 or the living creatures of Revelation 4, those forms serve symbolic purposes, not anatomical ones.

More to the point, many angels occupy a permanent station in the presence of God. The seraphim of Isaiah 6:2–3 stand before His throne, calling continuously, "Holy, holy, holy! The LORD of Hosts! His presence fills all the earth!" Gabriel, in Luke 1:19, identifies himself by saying, "I am Gabriel, who stands in the presence of God." The twenty-four elders of Revelation 4:10 fall before the throne in ceaseless worship. These beings do not marry, they do not reproduce, they do not require the biological mechanisms of a fallen world. They live in direct communion with their Creator, sustained not by food or lineage, but by proximity to the living God. This is precisely the state Jesus says the resurrected will share. As we will see, it is also the state that the renewed Adam of Genesis occupied before the fall.

Humans — Created, Made, or Formed?

No other aspect of creation enjoys the variety of descriptors that humans are granted, and that variety is deeply intentional. To begin with, humans were tasked with practical purpose: (Genesis 2:5b, JPS) "..to till the soil." More importantly, they were given what no other creature received—(2:7b) "the breath of life" from the very mouth of God who had spoken all creation into being.

In Genesis 1, we are told that God "created" [H1254] the heavens and the earth, yet He (1:7) "made" [H6213] the expanse, (1:16) "made" [H6213] the two great lights, (1:25) "made" [H6213] the wild beasts. And He said (1:26), "let Us make" [H6213] man—yet then (1:27), "God created" [H1254] man in His image.

The distinction is worth pausing on. The Hebrew word עָשָׂה (asah) [H6213], translated as made or make, is used in the broadest possible sense—to do, accomplish, appoint, advance, bestow. It covers the widest range of creative activity. But בָּרָא (bara) [H1254]—the same root as that second word of Genesis 1:1—is more precise. Strong's defines it as: to create (absolutely), and by extension, to cut down, select, feed (as formative processes). In other words, bara (create) carries the weight of something definitively chosen and brought forth, something selected out of and above everything else.

The creation account adds yet a third term in Genesis 2:7a, where God יָצַר (yatsar) [H3335]—formed—the human from dust. Strong's defines yatsar as: to mold into a form, especially as a potter shapes clay. This is the language of a craftsman working with intention and intimacy. God did not merely speak humanity into existence as He had the stars. He bent down, if we may speak in such terms, and shaped the human creature with a potter's care.

What emerges from these three terms—created, made, formed—is that humanity occupies a uniquely layered position in creation. We are simultaneously a product of God's broadest creative act and His most intimate personal attention. This matters enormously for understanding what the Adam was before the fall, and what the renewed Adam will be after the cycle completes.

The Adam, Who Was Not Yet Male

So then, what precisely was the condition of Adam in the garden? He was a created being who had neither need nor ability to reproduce. The Hebrew term הָאָדָם (ha'adam) [H120], as used in the opening chapters of Genesis, does not mean male human being. It means simply human being—and consistently in Genesis 2, the text refers to *the* adam with the definite article, meaning *the* human, or more precisely, the human one. The Hebrew grammatical form is masculine singular—but as with the angels, this reflects the conventions of the Hebrew language rather than an anatomical statement.

Adam did not become male until Eve was made. At that pivotal moment in Genesis 2:23–24, the male human is identified for the first time by the Hebrew term אִישׁ (Ish) [H376], meaning male, husband; and the female is called אִשָּׁה (Ishah) [H802], meaning female, wife. Prior to this moment, there is only ha'adam — *the* human. It is the arrival of Ishah that, by contrast and completion, makes Ish possible. Gender, in the Biblical narrative, is not the starting condition of humanity. It is something that emerges in response to a specific need. That need, as we will see, is directly tied to the fall.

After this point, Adam and Eve are capable of reproduction. But why introduce that capability at all? If God can simply form another human from the dust, why build into humanity the biological machinery of reproduction? The answer, viewed through the lens of the cycles, is deeply sobering: **reproduction is a prophecy of the fall!** God, in His foreknowledge, understood that this iteration of humanity would rebel and therefore required the ability to continue apart from Him. Biological reproduction is the mechanism by which a species survives in a world where walking face-to-face with God is no longer possible. From the moment of their conscious being, Adam and Eve were eternal creatures living in God's direct provision. Genesis 3 tells the story of how they traded that for independence—and death entered the world with it.

In this way, marriage and reproduction, sacred and beautiful as they are within the fallen world, are ultimately temporary institutions. They serve the fallen age. They will not survive into the renewed one.

The Remnant and the New Adam

This brings us to one of the most important—and frequently overlooked—threads running through the book of Revelation. Revelation does not end with the annihilation of all humanity. It ends with a remnant. Revelation 21:3 announces: "Behold, the tabernacle of God is with men, and He will dwell with them, and they shall be His people." The word men here is ἄνθρωποι (anthropoi) [G444]—human beings, generically, without gender specification. These are people who have passed through the fire of the apocalypse and arrived at the other side. They inhabit the New Jerusalem—the new Eden—where the river flows from the throne of God (Revelation 22:1), and the tree of life bears fruit every month (Revelation 22:2).

In the cyclical paradigm, this remnant is not an afterthought. It is the mechanism by which knowledge, memory, and awareness are carried forward from one cycle to the next. This is likely how recognition of the Precession of the Equinoxes—a cycle spanning 25,920 years—was preserved across civilizations that seemingly had no direct contact with one another. This is how awareness of the great ages—Golden, Silver, Bronze, Iron—persisted in the collective consciousness of cultures from Egypt to India to Mesoamerica. The remnant carries the memory. They are the thread that stitches one cycle to the next.

But there is something even more significant about this remnant. If Jesus's words in Matthew 22:30 are to be taken seriously—that the resurrected "neither marry nor are given in marriage, but are like the angels of God in heaven"—then the survivors of the apocalypse, those who inhabit the New Jerusalem, will have been transformed. They will no longer be Ish and Ishah. They will no longer be biologically gendered in the way that fallen humanity has been since Genesis 3. They will be, once again, ha'adam —*the* human, singular and complete, existing in direct communion with God, sustained by His presence rather than by biological reproduction.

This is the Adam of the next Genesis. Not a new creature invented from scratch, but a restored creature—humanity reset to its original design. The modifications required are not additions but subtractions: the removal of the biological scaffolding that was erected for a fallen world. What remains is what was always there beneath it—the eternal being, formed from the dust, animated by the breath of God, and made for nothing less than face-to-face fellowship with its Creator.

Yet here an apparent contradiction rises to meet us, and it deserves a plain answer. If a remnant is carried across the seam of the cycle, how can Genesis insist that the human was formed fresh from the dust of the ground? Either God shapes a new creature from the soil, or He preserves survivors from the age before—surely not both, and not both standing together in a single garden.

The contradiction is real only on the surface. Consider first what "formed from the dust" actually claims. The Hebrew yatsar pictures a potter working clay already in his hands; it is the language of shaping, not of manufacturing something from nothing. To be formed from the dust is the signature God sets on every human being — "for dust you are, and to dust you shall return" — not a tally of how many stood in Eden. And when Revelation declares, "Behold, I make all things new" (Revelation 21:5), the word is kainos [G2537]: not new in origin, as though the old were thrown away, but new in kind—the same creature, renewed.

The remnant are not a second batch of humanity set beside a freshly minted Adam. They are that Adam, re-formed.

Consider next what the resurrection works in them. Jesus is explicit: those counted worthy of that age "neither marry nor are given in marriage; nor can they die anymore, for they are equal to the angels and are sons of God, being sons of the resurrection" (Luke 20:35-36). The survivors of the apocalypse pass through precisely this change. They are returned to the state of ha'adam—singular, un-gendered, deathless, sustained by the presence of God rather than by lineage. Set down again upon the new earth as the cycle reopens, the feminine adamah once more yields the un-gendered human, exactly as in the first garden; and only afterward, as the new age bends again toward its fall, is Ishah drawn from the side of Ish, and the machinery of reproduction restored.

Notice the strange symmetry hidden in those two acts of making. At the garden's first moment the direction runs one way: ha'adam [H120], a grammatically masculine word, is drawn out of the adamah [H127], a grammatically feminine one—the masculine shaped from the feminine, lifted from the soil by the potter's hand of yatsar [H3335], to form. Then, when the woman is made, the direction reverses. She is not formed from dust at all but built—banah [H1129], the verb of a craftsman raising a structure—from the tsela [H6763], the side, of ha'adam. Now the feminine Ishah[H802] issues from the masculine Ish [H376]; source and offspring have traded places. The feminine gives rise to the masculine, then the masculine to the feminine, until the line bends back upon itself and closes like a ring. This is no accident of grammar. It is the motion of the cycle written in flesh and soil: what the end of one age unwinds—collapsing Ish and Ishah back into the single, ungendered ha'adam, the state the Greek of the resurrection calls isangeloi [G2465], "equal to the angels"—the dawn of the next age winds up again, carrying the pattern whole from one Genesis into the next.

Consider, finally, that Scripture never frames the renewed creation as a solitary pair. Revelation fills it with more than the Bride: there are "the nations" who walk by the city's light, and whose wounds the leaves of the tree of life are given to heal (Revelation 21:24; 22:2). Genesis does the same the instant we look past Eden's center to its edges. The newly formed Adam is the firstfruits—the head God shapes with His own hands to open the cycle—while the remnant are the nations gathered around him, carried forward from the age before. Both belong in the garden, because both texts, laid over one another as the single overlapping story they are, require them there: the created head and the carried-over multitude, side by side from the first page of the new age.

And it is that carried-over multitude—the wider humanity already present when the garden reopened—that the story's very next difficulty forces into the light.

Cain's Wife and the Depth of Pre-Fall Humanity

This framework also quietly resolves one of Scripture's most persistently puzzling questions. If Adam and Eve had only Cain and Abel, and Cain killed Abel—then where did Cain's wife come from? (Genesis 4:17) "Cain knew his wife, and she conceived and bore Enoch."

The answer, within the cyclical model, is straightforward: there were far more people on the earth by this time than the Genesis narrative explicitly accounts for. The remnant who survived the previous apocalypse would have been present —people who carried forward into this new cycle but who subsequently, and not surprisingly, rebelled against the God who had preserved them. Adam and Eve, viewed in this light, may not be merely two individuals but the representative story of all humanity—a résumé of the whole human experience compressed into a single narrative. They are, in the deepest sense, us. The story of the garden is not only their story. It is the story we have each lived in miniature.

And Eve herself was likely far more prolific than our Sunday school imaginations suggest. Genesis 3:16 records God telling Eve, "I will make most severe your pangs in childbearing; in pain shall you bear children." The severity of that curse implies a before—a prior state in which childbearing was not the ordeal it became. It is entirely plausible that Eve bore many children before the curse intensified that process. Humanity in its earlier, closer-to-original form was likely far more robust, far longer-lived, and far more capable than our diminished modern condition allows us to imagine.

The Circle Closes

What Genesis opens, Revelation closes—and then Genesis opens again. The ha'adam, that original ungendered human in intimate relationship with God, is not simply a character from the past. It is a prototype of the future. The New Jerusalem is the new Eden, and the new Eden will be populated by a new Adam—a remnant, transformed, no longer needing the institutions of marriage and reproduction because they will once again have what Adam and Eve surrendered: the unmediated presence of the living God.

"The Spirit and the bride say, 'Come!'" (Revelation 22:17). The invitation echoes across the cycle. The garden awaits. The river flows. The tree of life bears its fruit. And the human creature—finally, fully, and at long last—comes home.

2.6 Satan's Second Chance

If you are anywhere close to my age, you will remember one of the funniest comedians of the 1970s, Flip Wilson. One of Flip's most endearing recurring sketches featured his ongoing battles with the devil—a slippery, persuasive character who could talk anybody into doing something they knew perfectly well they shouldn't. After they had inevitably succumbed, the excuse was always the same: "The devil made me do it!"

Of course, that became one of the most convenient deflections in American popular culture. Underneath the laughter, the question it raises is genuinely serious. Is the devil always lurking—insidious, unseen, ever-present, and bent on our destruction? The short answer is - yes! The more complete answer, as we will see, is considerably more interesting than that.

The Jewish Understanding of the Adversary

Before we follow the serpent from the Garden of Eden all the way to the lake of fire in Revelation, it is important to understand what the ancient Hebrews actually thought about evil—because their perspective is considerably more nuanced than popular Christianity tends to assume.

In the Hebrew Bible, the term שָׂטָן (satan) [H7854] means, at its root: adversary or accuser. The Jewish Publication Society (JPS) Tanakh renders it consistently as "the adversary"—a functional title, not a proper name. This is a critical distinction. For the ancient Israelites, ha-satan — literally, *the* adversary—was not an independent force of cosmic evil operating in opposition to God. He was, if anything, more like a prosecuting attorney in the heavenly court, operating strictly within the bounds of what God permitted.

The clearest picture of this appears in Job 1:6–12, where ha-satan appears among "the divine beings" who present themselves before God. He is not crashing the gates of heaven—he belongs there, at least provisionally, and he cannot touch Job without explicit divine authorization. God limits his actions at every turn. The same functional adversary appears in Zechariah 3:1–2, accusing the high priest Joshua, only to be rebuked by the Angel of the LORD on God's behalf. Even the troubling passage in 1 Chronicles 21:1, where "Satan stood up against Israel and incited David," is illuminated by its parallel in 2 Samuel 24:1, where the same impulse is attributed directly to God. Hebrew thought never frames ha-satan as God's rival. He is more like God's instrument—a pressure applied to reveal what is already in the human heart.

This is why the ancient Israelites, when catastrophe struck, did not typically say "the devil did this to us." Their prophets were far more direct. Isaiah, Jeremiah, Amos—they looked their audiences in the eye and said, in effect: you did this to yourselves, and God is responding accordingly. Isaiah 45:7 puts it with characteristic bluntness: "I form light and create darkness, I make peace and create evil—I the LORD do all these things." God's sovereignty is total. The adversary operates within that sovereignty.

Rabbinic Judaism would later internalize this further. The Talmud (Bava Batra 16a) equates ha-satan, the yetzer hara (the evil inclination), and the Angel of Death as a single concept—not an external cosmic enemy, but the internal pull toward self-interest that makes genuine moral choice possible. Maimonides saw Satan not as a rebel angel but as a symbol of the force within the human soul that turns away from righteousness. For the Jewish mind, the real battlefield was always the human heart, not the cosmic heavens.

What Changes in the New Testament

The Christian reading of Scripture does not contradict this foundation so much as it extends it. Jesus clearly took the adversary with the utmost seriousness. In Luke 10:18, He tells His disciples, "I saw Satan fall like lightning from heaven"—a statement that implies both a prior elevation and a decisive witnessed fall. In John 8:44, speaking to the Pharisees, He identifies Satan as "a murderer from the beginning" who "does not stand in the truth, because there is no truth in him." These are not the words of someone treating the adversary as a mere metaphor for human weakness. Jesus engaged with Satan directly—in the wilderness temptation of Matthew 4, in the exorcisms recorded throughout the Gospels—and He consistently treated him as a real, active, and dangerous personal force.

The crucial development in the New Testament is the identification of the serpent of Genesis 3 with the dragon of Revelation. Revelation 12:9 makes this explicit: "So the great dragon was cast out, that serpent of old, called the Devil and Satan, who deceives the whole world." Four identities—dragon, ancient serpent, Devil, Satan—collapsed into one. This is the thread that runs from the Garden all the way to the apocalypse, and it is the thread that, within the cyclical framework of this book, becomes extraordinarily significant.

Bound for a Thousand Years — But When?

Here is where the cyclical reading of Scripture opens a genuinely provocative possibility.

Revelation 20:1–3 describes an angel coming down from heaven with "the key to the bottomless pit and a great chain in his hand." He seizes "the dragon, that serpent of old, who is the Devil and Satan" and binds him for a thousand years, "so that he should deceive the nations no more till the thousand years were finished. But after these things he must be released for a little while."

The standard Christian interpretation places this binding after the return of Christ described in Revelation 19—a future event, still ahead of us. And, within a strictly linear reading of Scripture, that makes sense. Within the cyclical framework this book has been developing—where the closing events of Revelation flow directly back into the opening events of Genesis—a different and deeply compelling reading emerges.

What if the thousand years of binding in Revelation 20 precedes the Garden narrative of Genesis 2 and 3? What if the "little while" of Satan's release, described in Revelation 20:3, corresponds precisely to the serpent's appearance in the Garden?

Consider the sequence within the cyclical model. The previous cycle ends in the catastrophic events of the apocalypse. Satan, the great adversary, is bound—contained, restrained, neutralized—as the new creation takes shape. The ha'adam of Genesis 2 lives in the Garden in unbroken fellowship with God. There is no hint of the adversary's presence during this period. The creation is declared "very good" (Genesis 1:31). God walks with His human creature. The river flows. The trees bear fruit. There is, as yet, no death.

And then — the serpent appears.

Genesis 3:1 introduces him with a word that should stop every careful reader in their tracks: נָחָשׁ (nachash) [H5175]—serpent. Strong's defines nachash not only as a literal serpent but, by extension, as one who practices divination, one who hisses or whispers. The root carries the idea of something that enchants—that works its influence through subtle, indirect means rather than direct confrontation. This is precisely how Revelation describes Satan's primary method of operation: "who deceives the whole world" (Revelation 12:9). He is not primarily a destroyer. He is primarily a deceiver.

That this particular creature appears in the Garden "more shrewd than any other beast of the field that the LORD God had made" (Genesis 3:1) is not incidental. It is a signature. The adversary who was bound has been released—"for a little while"—and he goes immediately to work doing what he has always done: whispering, questioning, deceiving. "Did God really say..?" (Genesis 3:1). The same voice that deceived the nations at the end of the previous cycle now turns its attention to the pristine remnant of the new one.

This reading resolves something that has long puzzled careful readers of Genesis. If this is the first creation, the first humanity, the first Eden—where does the serpent come from? Why is he already there, already shrewd, already in possession of a fully developed strategy for undermining the human relationship with God? Within a strictly linear model, this requires a separate backstory—a prior fall of Lucifer, an angelic rebellion in the heavenly realm—that Genesis never actually describes. Within the cyclical model, no such separate explanation is necessary. The serpent is there because he was released. His presence in the Garden is not a puzzle to be solved by importing material from outside the text. It is the direct fulfillment of Revelation 20:3—"he must be released for a little while."

The Pattern Beneath the Story

What the cyclical framework reveals is that Satan's role is not random or opportunistic. It is structural. He appears at the precise hinge point of the cycle—the moment when the new humanity, restored and renewed, is most vulnerable. His method is always the same: not brute force, but the quiet corruption of trust. He did not drag Eve to the tree. He asked her a question.

The Jewish instinct to locate moral responsibility in the human heart rather than in an external devil is not wrong — it is, in fact, confirmed by the Genesis narrative. God does not excuse Adam and Eve on the grounds that the serpent deceived them. They are still accountable. The yetzer hara, the inclination toward self-interest, was already present in them—and the serpent knew exactly how to find it. What the cyclical reading adds is simply this: the adversary's appearance in the Garden was not an accident of cosmic geography. It was the appointed "little while" of Revelation 20:3, the brief window between one binding and the next, in which the cycle's great test would be administered—and, as it has apparently been administered before, failed.

The good news, of course, is that the cycle does not end with the Garden. It ends—and begins again—with the Lamb.

2.7 What Jesus Knew and was Waiting For Us to See

There is a moment in the Gospel of John that tends to pass by too quickly. On the night of His arrest, speaking to His disciples in what John records as a long, intimate farewell, Jesus says something that should stop every careful reader cold. "I still have many things to say to you, but you cannot bear them now. However, when He, the Spirit of truth, has come, He will guide you into all truth." (John 16:12–13)

Read that again. Jesus—the one of whom John elsewhere writes, "In the beginning was the Word" (John 1:1)—explicitly tells His closest friends that there are things He has not yet told them. Not because He doesn't know them. Not because they are unimportant—but, because the disciples, in that moment, are not yet equipped to receive them. The implication is as remarkable as it is easily overlooked: the full understanding of what Jesus taught was never intended to arrive all at once. It was designed to unfold—over time, through the Spirit, in response to a world that would eventually be ready to hear it.

This is not a minor pastoral concession, it is a structural feature of how Jesus understood the relationship between divine revelation and human readiness, and it connects directly to one of the most intriguing and under-examined prophecies in the entire Old Testament.

Daniel: Sealed Book and the Promise to Increase the Knowledge

In the closing verses of the Book of Daniel, after the angel has delivered his final vision of the end times, he gives Daniel an instruction that is equal parts command and promise. Daniel 12:4 reads: "But you, Daniel, keep the words secret [H5640], and seal [H2856] the book until the time of the end. Many will range far and wide and (*the*) knowledge [H1847] will increase." Opening of the books and the unsealing is the theme of Revelation—but, *the* knowledge? The knowledge of what? Not knowledge in the broad amorphous sense of society accumulating general knowledge. But, specific knowledge—*the* knowledge.

In this case, I think the Hebrew Bible (JPS) illustrates the sense of the text better than the NKJV. Daniel is given insight, but told to secret it away for a later date. A later date when the readers would have a better developed ability to comprehend the significance of what Daniel could only glimpsed.

The Hebrew word rendered knowledge here is הַדָּעַת (ha'da'at, from דַּעַת da'at) [H1847]. Strong's defines it as: knowledge or awareness—and critically, it carries the connotation not merely of accumulated information but of intimate, relational knowing. It is from the same root used when Genesis says that Adam knew [H3045] his wife—a knowing that is personal, experiential, and transformative. When the angel tells Daniel that ha'da'at will increase at the time of the end, he is not predicting the internet or the library of congress. He is predicting something far more significant: a deepening of understanding—a new capacity to perceive what was always there in the Biblical text but could not previously be seen.

The book, Daniel is told, is to be sealed until the time of the end. Sealed—not destroyed, not lost, not rewritten. Preserved, intact, waiting for the moment when the seals could be broken and the meaning could finally emerge. The implication is that the Book of Daniel contains information that the generations immediately following Daniel could not fully understand. Not because it wasn't written clearly enough, but because the framework required to make sense of it hadn't yet developed in human consciousness.

We are living in precisely that moment. The framework—cyclical history, precessional time, the zodiacal narrative encoded in Daniel 7 and 8—is now available in ways it simply was not before. The convergence of archaeology, astronomy, ancient linguistics, and biblical scholarship that characterizes our generation is not coincidental. It is the unsealing.

And Jesus knew it was coming.

Jesus and Daniel: A Relationship the Gospels Make Explicit

It would be difficult to overstate how thoroughly Jesus was immersed in the Book of Daniel. The evidence is not subtle, it is right on the surface of the Gospel text for anyone willing to look.

The most obvious thread is the title Jesus used for Himself more than any other: Son of Man. The phrase appears in the Gospels approximately eighty times, always on Jesus's own lips, always as a self-designation. Its source is Daniel 7:13: "I saw in the night visions, and behold, with the clouds of heaven there came one like a son of man." This is not a coincidence of language. Jesus was deliberately, consistently, and publicly identifying Himself as the figure Daniel had seen coming to the Ancient of Days to receive an everlasting dominion. Every time

He said Son of Man, He was quoting Daniel to anyone with ears to hear.

The most explicit moment comes in Matthew 26:64, when the high priest demands that Jesus identify Himself under oath. Jesus's response is a direct conflation of Daniel 7:13 (NKJV)(and, Psalm 110:1): "You will see the Son of Man sitting at the right hand of the Power, and coming on the clouds of heaven." The high priest tears his robes! He understands exactly what Jesus is claiming—that He is the figure of Daniel's vision, and that His coming in judgment is not merely a future event but an already-inaugurated reality. This is not a casual biblical allusion, it is a precise theological claim, rooted in a precise reading of Daniel's night vision.

Then there is the Olivet Discourse of Matthew 24, which is, in many ways, Jesus's own commentary on Daniel. He references Daniel explicitly in Matthew 24:15: "Therefore when you see the 'abomination of desolation,' spoken of by Daniel the prophet, standing in the holy place (whoever reads, let him understand)." That parenthetical—whoever reads, let him understand—is extraordinary. Jesus is not simply citing a prophecy. He is issuing an invitation to deeper reading. He is saying, in effect: there is more here than the surface text reveals. Go back. Read carefully. Let the understanding come.

Seasonal Language and the Rhythm of Cycles

Jesus's engagement with cyclical themes goes well beyond His explicit references to Daniel. It runs through the very texture of how He taught.

The agricultural parables are the most obvious example. Jesus returns again and again to the imagery of planting, growing, and harvest—a vocabulary that is inherently cyclical. A seed goes into the ground. It dies. It rises. It bears fruit. The fruit is harvested. The field rests, and the cycle begins again. For an agrarian audience this was simply the language of daily life, but Jesus uses it to describe the Kingdom of God—suggesting that the Kingdom itself operates according to these same rhythmic principles of death, renewal, and harvest.

In John 4:35–38, speaking to His disciples near the well in Samaria, Jesus says something that carries a depth easily missed: "Do you not say, 'There are still four months and then comes the harvest'? Behold, I say to you, lift up your eyes and look at the fields, for they are already white for harvest!" And then, crucially: "For in this the saying is true: 'One sows and another reaps.' I sent you to reap that for which you have not labored; others have labored, and you have entered into their labors."

This is cyclical thinking made explicit. The disciples are reaping what others—the prophets, the patriarchs, Daniel himself—spent generations sowing. The harvest they are entering did not begin with them. They are participants in a story far longer than their own lifetimes, gathering fruit from seeds planted across centuries. The cycle of sowing and reaping, Jesus is saying, operates on a timescale that dwarfs any individual generation.

The parable of the wheat and tares in Matthew 13:24–30 adds another dimension. Both wheat and tares—the true and the counterfeit—grow together throughout the age, indistinguishable until the harvest. Only at the end of the age (Matthew 13:39) does the separation occur. Jesus is describing not just a future event but a pattern—the simultaneous growth of truth and deception through the long arc of history, resolved only at the cycle's culmination.

The fig tree parable of Matthew 24:32–34 is perhaps the most direct cyclical statement Jesus makes: "Now learn this parable from the fig tree: When its branch has already become tender and puts forth leaves, you know that summer is near." The fig tree doesn't announce summer. It signals it—through a natural process that repeats year after year, cycle after cycle. Jesus is telling His listeners to read the signs of the times the way a farmer reads the signs of the seasons. The signs are there. The cycle is operating. Those with eyes to see will recognize where in the cycle they stand.

Hidden Things and the Right Moment to Reveal Them

In Matthew 13:35, citing Psalm 78:2, Jesus explains His use of parables: "I will open My mouth in parables; I will utter things kept secret from the foundation of the world." Things kept secret from the foundation of the world. This is the language of deep concealment—not accidental obscurity but deliberate hiddenness, things placed out of reach until the appointed moment of their revealing.

This aligns precisely with the instruction given to Daniel: seal the book until the time of the end. Both passages share the same underlying theology—that divine revelation is not delivered all at once, in a single deposit, to be understood identically by every generation. It is layered, tiered, and timed. Certain understandings are withheld not because God is capricious but because the human capacity to receive them must first develop. The da'at (the knowledge) must increase before the sealed things can be opened.

Jesus understood this principle better than anyone. His use of parables was not just a teaching technique for simplifying complex ideas. It was a deliberate strategy of secreting-away and an invitation operating simultaneously—hiding the deeper meaning from those not yet ready while drawing the genuinely seeking reader further in. "He who has ears to hear, let him hear" (Matthew 11:15) is not a throwaway line. It is a standing invitation to a deeper level of engagement with the text than most readers attempt.

And then there is the remarkable statement of Matthew 11:25, where Jesus prays: "I thank You, Father, Lord of heaven and earth, that You have hidden these things from the wise and prudent and have revealed them to babes." The wise and prudent—those most confident in their existing interpretive frameworks—are precisely the ones who cannot see what the babes, the genuinely open and humble seekers, are able to perceive. The deepest readings of Scripture, Jesus is suggesting, will not come from those most heavily invested in the established understanding. It will come from those willing to look again, with fresh eyes and no predetermined conclusions.

What Jesus Was Waiting For Us to Understand

Drawing these threads together, a coherent picture emerges—one that this book has been building toward from its opening pages.

Jesus was not simply a teacher of personal ethics or a proclaimer of individual salvation, though He was certainly both of those things. He was a reader of the deep structure of history—of the cyclical, star-marked, divinely ordered rhythm of creation, fall, judgment, and renewal that Daniel had encoded in his visions and that the heavens themselves have been proclaiming since before human memory. He knew Daniel's framework intimately. He identified Himself explicitly with Daniel's "Son of Man". He taught in the cyclical language of sowing and harvest, of hidden things and appointed revelations, of seasons that signal what is coming to those who have learned to read them.

And He told His disciples plainly—there is more than you can currently bear. The Spirit will lead you into it. The knowledge will increase. The sealed book(s) will be opened.

Daniel 12:4 is not a prediction of technological progress. It is a promise that the generation living at the turn of the final cycle will be given the capacity to read Scripture in a way no previous generation could—because the full sweep of the cycle will, for the first time, be visible. The beginning and the end will be simultaneously in view. Revelation and Genesis will be recognizable as the two faces of the same turning wheel. And the zodiacal framework encoded in Daniel 7 and 8, dismissed for centuries as astrological contamination, will be recoverable as precisely what God always intended it to be: signs (Genesis 1:14)—divine markers charting the rhythm of history for every generation that has eyes to see them.

This is the knowledge that is increasing. And it is to Daniel's visions that we must now turn, with those newly opened eyes, to trace the complete arc of the cycle from its starlit beginning to its starlit end, and back again.

PART 3: Daniel, in the Nexus of Time

Daniel 1:20-21

20And in all matters of wisdom and understanding about
which the king examined them, he found them ten times
better than all the magicians and astrologers who were in
all his realm. 21Thus Daniel continued…

3.1 Daniel's Waypoints on the Path

Have you ever watched one of those nature documentaries where they follow a single river from its source high in the mountains all the way to the sea? At first its just a trickle—barely visible through the rock and snow. As it descends, other streams join it, valleys widen, and by the time you reach the delta you're looking at something vast and powerful that seems to have always existed. Understanding where you are in this book requires something similar. Before we enter Daniel's world, it is worth pausing to look back at the river we've been following—to see how the trickle became the current that is now carrying us into the most detailed and remarkable section of this journey.

A Summary of Where We Have Been

Part 1 of this book laid the essential groundwork for everything that follows. Its first concern was with how we read the Bible. The temptation—one we all feel—is to bring our own meaning to the text rather than drawing meaning out of it. We approach Scripture with conclusions already formed, shaped by tradition, by teachers we trust, by the communities we belong to. This book asks us to try something harder and more rewarding: to slow down, to sit with what the text actually says in its original languages, and to let the meaning emerge rather than imposing it. That discipline, applied consistently, is what makes the argument of this book possible.

From that starting point, Part 1 introduced the central proposition: that history, as God has arranged it, does not move in a straight line from creation to conclusion. It moves in cycles—grand, recurring arcs of creation, fall, judgment, and renewal. The evidence for this is not merely theological. It is encoded in the Precession of the Equinoxes—the 25,920-year wobble of the Earth on its axis that causes the twelve constellations of the zodiac to slowly rotate through the position of the spring sunrise over millennia. Ancient civilizations from Egypt to India to the Americas were aware of this cycle. Their monuments are aligned to it. Their mythologies encode it. The persistence of this knowledge across cultures that supposedly had no contact with each other suggests something more than coincidence. It suggests that this knowledge was preserved—deliberately carried forward from one cycle to the next by a remnant who survived whatever catastrophe closed the previous age.

Part 2 built directly on that foundation, and its implications were striking. The opening events of Genesis and the closing events of Revelation are not two endpoints of a single straight line. They are two descriptions of the same moment—the hinge point of the cycle, where one age ends and another begins. Simultaneously, the New Jerusalem of Revelation 21-22 is the Eden of Genesis 2, restored and perfected. The tree of life that appears in both is the same tree. The river that flows from the throne of God in Revelation 22 echoes the river of Genesis 2. God's tabernacling presence with His people, announced in Revelation 21:3, is precisely what the Garden represented before the fall.

Part 2 also addressed the most important human question the cyclical framework raises: what happens to people? If a remnant survives each apocalyptic transition, what does that remnant look like? The answer drawn from Jesus's own words in Matthew 22:30 is startling in its precision. The resurrected, He says, 'neither marry nor are given in marriage, but are like the angels of God in heaven.' The original Adam—the ha'adam of Genesis 2, before gender emerged through the creation of Eve—was such a being: non-gendered, immortal, walking in unmediated fellowship with God. The remnant who enters the New Jerusalem will be restored to that original condition. Marriage and biological reproduction, sacred institutions in the fallen world, served their purpose in the age of separation from God. In the renewed age, they are no longer needed.

The adversary's role in all of this was carefully examined as well. The ancient Hebrews understood ha-satan as an adversary or accuser—a functional role within God's sovereign court. But Part 2 proposed something that the cyclical framework makes newly visible: the serpent's appearance in the Garden of Eden was not a random intrusion. It was the fulfillment of Revelation 20:3's announcement that Satan 'must be released for a little while.' The thousand years of binding in Revelation 20 may well correspond to the period between the previous apocalypse and the serpent's arrival in Eden—the 'little while' of his release being precisely the window in which he deceived Eve and the cycle's great test was administered.

Finally, Part 2 established that Jesus Himself understood and taught within this framework. His consistent self-identification as the 'Son of Man'—drawn directly from Daniel 7:13—was not incidental. It was a precise theological claim, tying His own identity to the zodiacal and cyclical vision of Daniel's night visions. His teaching in parables used the inherently cyclical language of sowing, growing, and harvest. And His promise in John 16:12-13—that there are things He has not yet told us, things the

Spirit will guide us into—opened the door to exactly the kind of deepening understanding that Daniel 12:4 promised: that *the* knowledge would increase at the time of the end.

That is where we stand as we enter Part 3. We have the cyclical framework. We have the Genesis-to-Revelation bookend structure. We have the transformed Adam, the bound and released adversary, and Jesus's own reading of Daniel. Now it is time to meet the man whose visions provided the backbone of this entire argument—the man who stood at one of the most significant nexus points in the history of human civilization.

Daniel an Introduction — The Person, The Visionary

There is a question that tends to catch people off guard when they first encounter it. If you ask most Christians who the great prophets of the Old Testament were, you would likely hear Isaiah, Jeremiah, Ezekiel, Amos, Hosea—possibly Micah or Zechariah. Daniel almost certainly appears on that list. He has a whole book of the Bible named after him. He interprets dreams for kings. He sees visions of the future that are still being argued over by scholars today. Surely he is a prophet, but here is the surprising thing, Jewish tradition does not consider him to be such.

In the Hebrew Bible—the Tanakh—the books are organized into three major sections. The Torah (the five books of Moses) comes first, followed by the Nevi'im (the Prophets), and then the Kethuvim (the Writings). The prophetic books—Isaiah, Jeremiah, Ezekiel, and the twelve minor prophets—are all in the Nevi'im. But Daniel? Daniel is in the Kethuvim, alongside Psalms, Proverbs, Job, and the historical books of Esther, Ezra, Nehemiah and Chronicles.

This classification is not an oversight. The distinction matters. The Hebrew prophets typically operated in a very specific mode: they received a word from God and delivered it directly to Israel, usually in the second person—'Thus says the LORD...' They spoke to their own generation, addressing specific sins and specific consequences, with one eye on the immediate future and another on the distant messianic horizon. Daniel does something different. He interprets the dreams of foreign kings. He has visions of sweeping historical epochs that unfold across centuries. He writes in the first person, describing what he personally experienced and saw. His framework is not primarily Israel's covenant relationship with God but the great sweep of Gentile world history—what he calls 'the times of the Gentiles.' It is history, seen from above, written for the long term.

This distinction between Daniel-as-historian and Daniel-as-prophet is actually quite useful for our purposes. It means that when Daniel describes the beasts of his night visions, he is not primarily issuing a warning to his own generation. He is recording what he sees in the heavens—the great constellations that mark the sweep of civilizations—and labeling them with the empires they represent. He is a chronicler of cycles, writing in the star-language that God established at creation when He said, 'Let there be lights in the expanse of the sky to separate the day from the night; they shall serve as signs' (Genesis 1:14). Daniel reads those signs and writes down what they say.

Babylonian Exile: The Arithmetic of the Sabbath

Before we can appreciate the significance of Daniel's arrival in Babylon, we need to understand why the Jews were there in the first place. The answer is both practical and theological—and it involves a piece of arithmetic that few people notice.

God's instructions to Israel regarding the land were remarkably specific. Every seventh year, the land was to rest. No sowing, no pruning, no harvesting. The ground would lie fallow for an entire year — what Leviticus 25:4 calls 'a sabbath of the LORD.' The purpose was multiple: rest for the soil, provision for the poor who could eat whatever grew wild, and, most importantly, a declaration that the land ultimately belonged to God and not to its human cultivators. 'The land shall observe a sabbath of the LORD' (Leviticus 25:2). It was not optional.

The additional instructions were practical and reassuring. God anticipated the obvious anxiety: 'And if you ask: What are we to eat in the seventh year, if we may neither sow nor gather in our crops?' (Leviticus 25:20). His answer was a promise of supernatural abundance in the sixth year—enough to carry Israel through the seventh year and into the eighth until a new harvest could come in. The system was elegant and entirely dependent on trust in God's provision.

Exodus 23:10-11 states the principle concisely: 'Six years you shall sow your land and gather in its yield; but in the seventh year you shall let it rest and lie fallow.' And Deuteronomy 15:1-2 extends the same seven-year rhythm into the cancellation of debts—the shemitah, שְׁמִטָּה [H8059], meaning 'a release.'

Israel largely ignored all of it.

For century after century, the sabbath years came and went unobserved. The land was worked continuously. The debts were not released. The seven-year rhythm that God had built into the covenant life of the nation was simply set aside in favor of economic productivity. The consequence, when it finally came, had a precise mathematical quality that should give any skeptic pause.

II Chronicles 36:21 explains the Babylonian exile with an exactness that is almost clinical: 'to fulfill the word of the LORD spoken by Jeremiah, until the land paid back its sabbaths; as long as it lay desolate it kept sabbath, to complete seventy years.'

Seventy sabbath years. Seventy missed observances, at one every seven years, equals 490 years of continuous violation. Four hundred and ninety years during which the land never rested as God had commanded. Now, in the forced absence of the people, the land would finally get what it was owed. The Babylonian exile was not arbitrary punishment—it was the land collecting its debt. And the precision of that calculation is itself a kind of theological statement: God keeps accounts, and He keeps them exactly.

It is into this moment of divine accounting—the end of 70 years, the beginning of a new possibility—that Daniel arrives. Or rather, it is into the beginning of those 70 years that he arrives as a young man, destined to live through the full duration of the exile and to stand at its conclusion when Cyrus the Great issued his edict allowing the Jews to return to their land.

Daniel and Joseph: Two Men at Two Pivotal Moments

It is impossible to read the story of Daniel in Babylon without being reminded of Joseph in Egypt. The parallels are too numerous and too precise to be accidental. Both were young men of exceptional quality taken from their homeland under duress. Both were placed in foreign courts where they rose to positions of extraordinary influence through the divine gift of interpreting dreams. Both served multiple rulers across the span of their careers. Both used their position not merely for personal advancement but to preserve and serve their people and their God. And, both found themselves at the precise historical moment when something foundational was being established for the future of Israel.

In Joseph's case, as Part 2 of this book explored in detail, that foundational contribution was the development of Old-Hebrew—the phonetic writing system that would become the basis for virtually every alphabet used in the Western world. Joseph's privileged position in Pharaoh's court, combined with his deep familiarity with both Egyptian hieroglyphics and the Semitic languages of his homeland, put him in a unique position to develop a simplified phonetic script that ordinary people could learn and use. The mining inscriptions of the Sinai peninsula—bearing Old-Hebrew characters alongside hieroglyphics of the same age—are the physical evidence of that linguistic revolution. See Inscription 2.1, in section 2.3 The Foundations of Written Hebrew and the Bible.

By the time Daniel arrived in Babylon, that Old-Hebrew script had been in use among the Israelites for centuries. Moses had written the Torah in it. It had been refined and developed through the long centuries of Israel's life in Canaan. But, Daniel's Babylon was about to change that. And Daniel's position at the center of Babylonian intellectual life put him, once again, at the precise moment when a foundational transformation was taking place —this time in the written form of the Hebrew language itself.

A Man at the Nexus

The Hebrew word דָּנִיֵּאל (Daniyyel) [H1840] means, simply, 'God is my judge.' It is a name that suits the man perfectly. In everything Daniel does—from his refusal to eat the king's food in chapter 1, to his continued practice of prayer when it had been made a capital offense in chapter 6, to his candid interpretation of visions that no one else dared to explain—Daniel operates as a man whose primary allegiance is to a court higher than any earthly throne. He was not defiant for the sake of defiance. He was not a zealot. He was, in the most precise sense, a man of principle who had decided very early in life what he was willing to die for and what he was not, and who stuck to that decision with remarkable consistency over a very long career.

That career spanned approximately seventy years—from the first deportation of Judah by Nebuchadnezzar in roughly 605 BC to the early years of Cyrus the Great after 536 BC. During that time Daniel served at least three major powers: the Babylonian Empire under Nebuchadnezzar and his successors, the brief reign of the Median king Darius (Daniel 5:31), and the early Persian Empire under Cyrus. Each transition brought new dangers and new opportunities, and Daniel navigated them all with a combination of diplomatic skill, genuine faithfulness, and an obvious intelligence that his contemporaries found—somewhat unnerving.

Daniel 1:20 notes that in all matters of wisdom and understanding, Nebuchadnezzar 'found them ten times better than all the magicians and astrologers in all his realm.' The Hebrew term for 'magicians' here is חַרְטֻמִּים (chartummim) [H2748], referring to the scribal scholars and dream-interpreters of the Babylonian court. The 'astrologers' are the אַשָּׁפִים (ashshaphim) [H825]—exorcists and enchanters. Daniel was not competing with amateurs. He was surpassing the intellectual elite of the most sophisticated civilization on earth at that time. And his superiority was attributed not to better training but to God's gift of wisdom.

When Nebuchadnezzar promoted him after the interpretation of his dream in chapter 2, Daniel was made 'ruler over the whole province of Babylon and chief prefect over all the wise men of Babylon' (Daniel 2:48). This is not a minor administrative appointment. The wise men of Babylon—the chartummim, the astrologers, the interpreters, the priests—were the intellectual and religious governing class of the empire. They maintained the libraries and star records. They trained new scholars. They served as advisors to the king on every question of policy, religion, and natural phenomena. Daniel was made their chief.

What this means, in practical terms, is that Daniel had direct authority over three of the most significant intellectual developments of his era—the three things we are calling landmark events in this section. He was positioned to influence the codification of Hebrew writing, the systematization of the zodiacal calendar, and the emerging religious movement that would eventually become Zoroastrianism. He did not necessarily originate any of these things. But his presence, his authority, and his theological perspective left fingerprints on all three.

Landmark Event 1: The Codification of Hebrew Block Script

If you have ever seen a page of a traditional Hebrew Torah scroll, you have encountered one of the most immediately recognizable writing systems in the world. The square, block-form letters—angular, precise, each character balanced within an invisible grid—have a quality of permanence about them. They look like they were designed to last. And in a sense, they were—because the form of those letters was deliberately standardized during the period of the Babylonian exile, at exactly the time when Daniel was exercising his authority over the intellectual life of Babylon.

To understand what happened, we need to go back briefly to what Part 2 established about the origins of Hebrew writing. Joseph, in Egypt around 1880BC, developed a phonetic script adapted from Egyptian hieroglyphics that allowed Semitic speakers to write their language in a simplified, accessible alphabet. This proto-Sinaitic script—the evidence for which is carved into the walls of the Sinai mining regions—became the foundation for what we know as Old-Hebrew, the script in which Moses wrote the Torah. It had angular pictographic features that connected it visually to its ancient origins.

By the time of the Babylonian exile, however, the Jews were being immersed in a culture that had its own highly sophisticated writing tradition. Babylonian Aramaic—the lingua franca of the Near Eastern world—was written in a script that had evolved from the same Phoenician root as Old-Hebrew, but had taken a different developmental path. The Aramaic script was more fluid, more rounded, and increasingly standardized through the bureaucratic demands of the Babylonian and later Persian empires. It was, in many ways, a more practical writing system for daily administrative use.

The Jews absorbed it rapidly. Daniel 2:4 notes that the Chaldeans began speaking to Nebuchadnezzar 'in Aramaic' — and then the text of Daniel itself switches to Aramaic and stays there through chapter 7. Daniel and Ezra are the only books of the Hebrew Bible to contain extended Aramaic sections. This is not coincidental. It is a linguistic fingerprint of the exile period, evidence that the Jewish scholars living in Babylon were operating in a bilingual environment where Aramaic was the language of power and daily commerce.

What emerged from this immersion was the gradual transformation of the Hebrew written form. The square, block-form letters—known in rabbinic literature as Ktav Ashuri, meaning 'Assyrian script'—were not invented from scratch. They were the result of the Hebrew writing tradition absorbing the more regularized forms of the Aramaic script and developing from that contact a distinctly Hebrew block form. The Talmud, in tractate Sanhedrin 21b-22a, attributes the codification of this script to Ezra the Scribe, who returned from Babylon to Jerusalem around 458 BC—a generation after Daniel's time, but building directly on the intellectual work done during the exile.

The Babylonians had a passion for systematization that made this kind of codification possible and even inevitable. It was a characteristic of Babylonian intellectual culture to take things that existed in multiple variant forms and reduce them to a canonical standard (more will be said about this in the next section). They did this with law (Hammurabi's Code), with astronomy (the MUL.APIN star catalogs), with weights and measures, and with language itself. The Aramaic script that dominated Babylon during Daniel's era had already undergone exactly this kind of standardization. The contact between the Jewish scribal tradition and the Babylonian love of canonical precision produced, over the course of the exile, something new: a form of Hebrew writing that was more regular, more standardized, and more legible than its Old-Hebrew predecessor.

Daniel's role in this process was almost certainly more than passive. As chief of the wise men of Babylon, he presided over precisely the scribal establishment that maintained and developed these writing traditions. The Babylonian scribal schools—the edubba were the repositories of astronomical records, legal tablets, literary texts, and linguistic standards. Daniel had authority over the scholars who worked in these institutions. And Daniel, as a man deeply committed to the preservation of his people's identity and traditions, would have had every reason to ensure that the Hebrew writing system was maintained, refined, and standardized during the long decades of exile when the risk of cultural dissolution was very real.

The physical evidence of this transition is visible in the Dead Sea Scrolls, discovered at Qumran in 1947. Most of the scrolls are written in the square block-form Hebrew that became standard after the exile. But some of the oldest scrolls, and some sections within later scrolls, still use the Old-Hebrew script for one specific purpose: writing the divine name—the Tetragrammaton (𐤉𐤄𐤅𐤄) YHWH (remember, Hebrew is written right-to-left). [For reference, the Tetragrammaton in block-form Hebrew looks like this - יהוה.] Scribes who had adopted the new block script still reverted to the ancient Old-Hebrew form when writing the name of God—a practice that speaks to the deep conservatism of the scribal tradition even as it was being modernized. This detail is itself a kind of bridge between the old writing and the new, a remnant of the past preserved within the transformed present.

The significance of this for the Bible's long-term transmission cannot be overstated. The block Hebrew script that was standardized during and after the Babylonian exile became the permanent form in which the Hebrew Bible would be copied, studied, and preserved for the next two and a half millennia. Every Torah scroll produced since then, every Talmudic manuscript, every medieval Hebrew Bible—all of them use this script. The work done in Babylon, in the intellectual environment over which

Daniel presided, established the written form of the Hebrew language for all time.

Joseph gave Israel its first alphabet. Daniel gave it its final form.

Landmark Event 2: The Babylonian Zodiac — Organizing the Heavens

To a modern reader, the word 'zodiac' carries the slightly embarrassing association of newspaper horoscopes and personality types determined by birth month. This book has no interest in any of that, and neither did the Babylonians who first systematized the zodiac in its most recognizable form. For them—as for Daniel, as for the purposes of this book—the zodiac was something far more serious: it was the first science.

Long before Babylon, the peoples of Mesopotamia had been watching the stars. The Sumerians maintained star records as early as 3000 BC. The Akkadian empire that followed them continued and expanded this tradition. It was during the later Babylonian period—roughly 1000 to 400 BC, with major developments occurring in the 7th and 6th centuries BC, precisely the era of Daniel—that Babylonian astronomy became something qualitatively different from anything that had come before. It became systematic. It became mathematical. It became, for the first time, genuinely predictive.

The key document in this transformation is the MUL.APIN series—a collection of cuneiform tablets whose name means 'Plow Star,' referring to the first constellation listed. These tablets, which exist in copies dating from around 700 BC, represent the synthesis of centuries of accumulated Babylonian astronomical observation. They catalog the positions of stars, the risings and settings of constellations, the paths of the moon and planets, and the relationships between celestial events and calendar dates. They are, in essence, the operating manual for Babylonian astronomy.

Embedded within the MUL.APIN tradition was the concept that would eventually become the zodiac as we know it: the recognition that the sun, moon, and planets all move through a relatively narrow band of sky—the ecliptic—and that this band could be divided into twelve sections, each corresponding to a constellation visible during a particular month of the year. By the 5th century BC, the Babylonians had codified these twelve sections into what scholars call the Babylonian celestial zodiac: twelve constellations of roughly equal extent, arranged in sequence around the ecliptic, used to track and predict the positions of celestial bodies.

The significance of this for Daniel's visions in chapters 7 and 8 is direct. When Daniel sees a lion, a bear, a leopard, and a horned beast arising from the great sea, he is—this book argues—seeing the great constellations of the ecliptic band, and some standard constellations just off-the-ecliptic, in their sequential order. See section 3.6 Cancer, Gemini and Off-the-Ecliptic to understand the significance of those constellations. The celestial imagery he uses was precisely the imagery that the Babylonian astronomical tradition had been organizing and codifying during the decades immediately before and during his career. He was not inventing a new symbolic language. He was using, for theological purposes, a visual vocabulary that was being refined and systematized in the very institution over which he held authority.

The Sexagesimal System

Among the Babylonians' most enduring contributions to human civilization is a mathematical system that most people use every single day without ever thinking about its ancient origins. It is called the sexagesimal system—a base-60 number system—and it underlies virtually every measurement of time and angle that modern civilization depends on.

The reason for this is simple: 60 is an extraordinarily useful number. It is divisible by 2, 3, 4, 5, 6, 10, 12, 15, 20, and 30. This means that fractions of 60—halves, thirds, quarters, fifths, sixths—all come out as whole numbers. No messy decimals. No remainders. For a civilization doing practical mathematics without a calculator, this property was enormously valuable.

Carefully hidden within the text of Daniel 1:20 is a fascinating contradiction to the sexagesimal system. Rooted within the sexagesimal is the number 6—and, 60 is clearly a multiple of 6. Why is this important!? Daniel 1:20 does not actually say, "..in all matters..they (the Jewish kids) were found to be ten times better.." In Hebrew it says, "..they were found to be a ten-hand better.."! What does that imply, or even mean? I will keep this one close to the vest for now and say only—this is a major theme in the next book in this collection and it may be of more importance than you could ever guess!

From the 360-degree circle, the Babylonians also developed the system of dividing each degree into 60 minutes of arc, and each minute into 60 seconds of arc. They applied the same logic to time: one hour equals 60 minutes, one minute equals 60 seconds. These conventions, established in Babylon more than two and a half millennia ago, are still in use on every clock face, every protractor, and every GPS coordinate in the world today. When you set your watch, you are using a Babylonian invention.

The ecliptic coordinate system that the Babylonians developed for mapping the zodiac was built on these mathematical foundations. By establishing the ecliptic as a reference circle and dividing it into 360 degrees, the Babylonians created a framework within which the positions of stars and planets could be described with mathematical precision and communicated clearly across time and geography. This was astronomy's first universal language—and its coordinates are still in use in modified form in modern celestial mechanics.

Daniel, as the chief overseer of Babylon's scholarly establishment, presided over the very institution that was refining and applying these mathematical tools. The astronomical records and calculations being produced in the Babylonian scribal schools during his tenure were among the most sophisticated intellectual work being done anywhere in the world at that time. His access to and authority over this tradition would have been complete.

Babylon as Organizer of the World

The impulse to systematize—to take disparate, variant forms of anything and reduce them to a clean, canonical standard—was not limited in Babylon to astronomy and mathematics. It was a pervasive cultural characteristic, visible across virtually every dimension of Babylonian life.

The city of Babylon itself was one of the ancient world's most deliberately planned urban environments. The city was laid out with a regularity that reflects the same love of order visible in Babylonian science. The Processional Way—the great ceremonial boulevard known in Babylonian as Aibur-shabu—ran in a straight line through the heart of the city, lined with glazed brick walls decorated with lions and dragons, from the Ishtar Gate to the temple complex of Marduk. This was urban design as theology—the orderly city as a reflection of divine order.

Hammurabi's Code, dating from around 1754 BC but representing a tradition that continued into later Babylonian culture, was one of the ancient world's most comprehensive attempts to codify law—to take the accumulated customs and case-law of centuries and arrange them into a systematic, public, and consistently applied legal framework. The stele on which the code was inscribed was set up in public places so that any citizen could read it. Order, in Babylon, was meant to be visible and accessible.

Babylonian canal and irrigation engineering represented the same systematic mindset applied to the physical landscape. The agricultural productivity of Mesopotamia was almost entirely dependent on managed water—canals dug from the Euphrates and Tigris to water the surrounding fields. These systems were maintained, extended, and documented with a precision that allowed the same canals to remain functional across centuries.

Weight and measure standards were another area of systematic Babylonian organization. The mina, the shekel, and the talent — units of weight that appear throughout the Hebrew Bible—originated in the Babylonian standardization of commercial measures. These standards were enforced by the state and backed by official weights kept in temples as reference standards. Trade across the empire depended on this kind of systematic consistency.

It is in this context that the Babylonian systematization of the zodiac makes complete sense. It was not an isolated intellectual achievement. It was the expression of a culture-wide impulse to organize the important things—to bring heaven into the same kind of orderly precision that Babylon had brought to law, commerce, water, and urban space. Daniel was not just tolerated within this culture, he thrived in it, because the

Hebrew tradition of careful, ordered reading of God's creation resonated with the Babylonian love of systematic observation. Where they differed was in the ultimate source of order: for Babylon, it was the gods of the pantheon; for Daniel, it was the one God who made everything.

The origins of the constellation forms themselves—the specific images assigned to the twelve ecliptic sections—likely predate Babylon significantly. The basic iconography of the bull, the lion, the scorpion, and the water-bearer appears in Sumerian art and mythology going back to at least 3000 BC. Some scholars have traced certain constellation images to an even more ancient source, arguing that the original set of stellar iconography may have been established when those constellations occupied positions in the sky that were astronomically significant—positions consistent with a period around 4000-6000 BC, before the Precession of the Equinoxes had shifted them from their original alignment. The cycle that this book has been tracking, in other words, may have left its mark on the very constellation images that Daniel uses to describe it.

Landmark Event 3: Daniel and the Roots of Zoroastrianism

Of the three landmark events associated with Daniel's time in Babylon, this one is simultaneously the most intriguing and the most difficult to document with certainty. The historical record is simply less clear here than it is for the development of Hebrew script or Babylonian astronomy. The circumstantial evidence is compelling enough—and the theological implications are significant enough—that it deserves careful examination.

The question is this: did Daniel play a role in the early development of the religious movement that is now Zoroastrianism? And if so, what does that connection tell us about the extraordinary event recounted in Matthew 2, when a group of 'wise men (magi) from the east' traveled hundreds of miles to find an infant Jewish Messiah?

The Problem of Dating Zoroaster

Zoroastrianism takes its name from its founder—or, more precisely, from the Greek rendering of his name. In the Avestan language of ancient Iran, he is called Zarathustra. His date is one of the most persistently disputed questions in the history of religion. Traditional Zoroastrian sources place him around 600 BC—which would make him a near-contemporary of Daniel. Greek sources from antiquity placed him thousands of years earlier, which is almost certainly legendary exaggeration. Modern scholars working from the internal evidence of his oldest writings — the Gathas, a collection of hymns attributed to Zarathustra himself—generally place him somewhere between 1500 BC and 600 BC, with many favoring a date in the range of 1000-700 BC.

The uncertainty matters for our purposes because it shapes what we can say about Daniel's possible influence. If Zoroaster lived long before Daniel, then Daniel's contribution would have been to an already established religious tradition rather than to its founding. If the traditional 600 BC dating is correct, the two were essentially contemporaries, and the influence could have been more direct and fundamental. Most likely, what Daniel encountered in Babylon was an emerging religious tradition that had roots going back centuries to Zarathustra's original teachings. Teachings that were in the process of being systematized and institutionalized in a form that the Persian ruling class would adopt as their official religion.

What is historically clear is this: by the time of Cyrus the Great, who conquered Babylon in 539 BC during Daniel's lifetime, the Persian ruling class was

Zoroastrian in orientation. Cyrus himself demonstrated a theological tolerance that some scholars connect to Zoroastrian universalism—the idea that the divine truth is accessible to all peoples, not only to Persians. His famous edict allowing the Jews to return to their homeland and rebuild their temple, recorded in Ezra 1:1-4, is one of the most remarkable acts of religious tolerance in the ancient world, and it is entirely consistent with a ruler who believed in a single supreme deity of truth and light rather than a tribal god with no interest in other peoples.

Daniel's Choice: Religion Over Administration

The Book of Daniel contains several passages that, taken together, reveal something important about the nature of Daniel's position in Babylon. After Nebuchadnezzar appointed him chief prefect over all the wise men in Daniel 2:48, something interesting happens in Daniel 2:49: 'Daniel made a request of the king and he appointed Shadrach, Meshach, and Abed-nego over the affairs of the various provinces of Babylon, while Daniel himself remained at the king's court.'

Read carefully, this verse suggests that Daniel arranged for his three friends to handle the day-to-day administrative work in the provinces while he himself remained at the king's court in what appears to have been a more direct advisory and interpretive role. Later, in Daniel 6:1-3, under the Median king Darius, Daniel is appointed as one of three presidents over the 120 satraps of the kingdom, with the king specifically intending to 'set him over the whole kingdom.'

The chapters that follow are dominated not by Daniel's administrative work but by his prayer practices, his prophetic visions, and his faithfulness to his God—the things that got him thrown into the lion's den. The picture that emerges is of a man who held significant official authority but who consistently oriented that authority toward theological and prophetic concerns rather than pure administration. He was not primarily a bureaucrat who happened to have religious convictions. He was primarily a man of God who had been given governmental authority as a platform for God's purposes. The most significant of those purposes, from the perspective of this book, was his ability to shape the religious understanding of the Babylonian and Persian priestly class—the magi.

Daniel and the Magi

The word rendered 'wise men' in most English translations of Matthew 2:1 is the Greek term μάγοι (magoi) [G3097]. This is the same word from which we get the English words 'magic' and 'magician,' but its original meaning was far more specific. The Magi were the priestly and scholarly class of ancient Persia — the religious intelligentsia who maintained the sacred fire, interpreted omens, advised kings, and preserved the theological traditions of their people. By the time of Jesus's birth, they were associated with Zoroastrianism, the official religion of the Persian and Parthian empires.

The connection between Daniel and the Magi is explicit in the Book of Daniel itself. When Nebuchadnezzar demanded that his wise men both tell him his dream and interpret it—an impossible task, which Daniel accomplished by divine revelation—the king promoted Daniel to 'chief of the magicians, enchanters, Chaldeans, and diviners' (Daniel 5:11). Daniel was not merely associated with the magi, he was their leader.

In that position of leadership, over the course of his nearly seventy-year career, Daniel had both the opportunity and the motivation to share his theological convictions with the priestly class he oversaw. His convictions included the belief in one sovereign God, the creator of heaven and earth. They included the expectation of a coming Messiah —'one like the Son of Man' who would receive 'an everlasting dominion' (Daniel 7:13-14). They included a cosmology in which the heavens were filled with meaningful signs—divine markers that could be read by those who knew how.

The Zoroastrian tradition that the Magi of Matthew 2 represented had absorbed, over the preceding five centuries, a strikingly similar set of convictions. Zoroastrianism taught the sovereignty of Ahura Mazda—the 'Lord of Wisdom'—as the supreme divine being. It taught the coming of a world-savior figure called the Saoshyant, born of a virgin, who would defeat evil and bring about the renewal of creation. It taught resurrection of the dead and a final judgment. It placed enormous importance on the stars as divine signs.

The similarities between these two theological traditions—Hebrew monotheism and Zoroastrian eschatology—are too numerous and too precise to be entirely coincidental. The most natural explanation for them is contact and influence. And the most natural vehicle for that contact and influence is Daniel, who for most of seventy years led the very scholarly and priestly institution through which Zoroastrianism was developing its theological vocabulary and institutional structure.

Early in my research for this book, as I initiated my study of Zoroastrianism, I was struck by the thought that Daniel might actually be Zoroaster (Zarathustra). In fact, the uncertainty in dating the existence of Zoroaster himself made the thought even more compelling. In the end, I'm still not entirely sure—I might be correct! Table 3.1 (below) is a list of similarities and differences between Zoroastrianism and Judaism in the time of Daniel's influence.

The Magi who came to Bethlehem were the inheritors of a tradition that Daniel had shaped. They came from the east—from Persia (modern Iran), the heartland of Zoroastrianism—carrying an expectation of a Jewish Messiah-king that they had preserved for over five hundred years. They knew to look for a star because Daniel had told their predecessors to look for a star. They knew it would signal a royal birth because Daniel had told them it would. And when the star appeared, they knew what to do.

The star itself appeared in the east—not in the direction of Bethlehem, but rising in the sky above Persia, signaling that the moment had come. Once in Jerusalem, the Magi had to consult the Jewish scriptures to locate the birthplace precisely. The star was an announcement, not a navigation system.

Micah 5:2 provided the address: “And you, O Bethlehem of Ephrath, least among the clans of Judah, from you one shall come forth to rule Israel for Me—One whose origin is from of old, from ancient times.”

The theological chain from Daniel's Babylon to the birth of Jesus in Bethlehem, as traced through the Magi, is one of history's most remarkable threads of divine Providence—stretching across five centuries, multiple empires, and two religious traditions, yet arriving with the precision of a completed appointment.

Table 3.1: Judaism & Zoroastrianism

Topic	Zoroastrian	Jewish	*
Supreme God	Ahura Mazda	YHWH	S
Monotheism	Evil near-equal to God	Evil subordinate to God	D
Dualism	Co-equal cosmic struggle	Subordinate evil forces	~S
Angels	Holy immortal hierarchy	Named archangels (Daniel)	S
End Times	Cosmic renewal/ renovation	Final judgment (Daniel)	S
Resurrection	Core doctrine	Emerges in Daniel 12	S
Sacred Symbol	Fire & light	Fire & light	S
Ethics	Truth vs. lie	Righteousness vs. evil	S
God & History	God rules all empires	Same — central in Daniel	S
Law	No covenant law	Torah covenant	D
God's People	Universal	Israel specific	D
Worship	Fire temple, no sacrifice	Temple & sacrifice	D
Purity	Elements & corpse avoidance	Food, ritual, body	D
Future Savior	Saoshyant	Messiah	~S
Context	Persian court religion	Daniel in Persian court	—

* S = Similar · D = Different · ~S = Partial

From the Nexus to the Visions

Daniel arrived in Babylon as a young exile with a gift from God and a commitment to the God who gave it. Seventy years later, still at his post—having shaped three of the most consequential intellectual developments of the ancient world - the Hebrew alphabet found its permanent form, the Babylonian zodiac was systematized into the star-language that Daniel himself would use for prophecy, and the priestly class who would one day travel to Bethlehem to find a Jewish Messiah received their instructions.

Now it is time to turn to the visions themselves—the night visions of Daniel chapters 7 and 8, in which the great constellations unfold as a map of human history. Daniel did not just stand at the nexus of his time. He was given the eyes to see what that time pointed toward—the long arc of empires rising and falling, the coming of the Lamb, and this final judgment that would close the cycle and open the next one. What follows is that vision, read in the star-language that Daniel and his God established together.

3.2 Daniel 7: The Cycle Revealed

Most people assume they have to be a scholar, or at least a particularly devoted student of Scripture, to make sense of Daniel 7. The chapter has a reputation for being dense, symbolic, and frankly a little frightening. Four terrible beasts arising from a churning sea. A court of divine judgment. An 'Ancient of Days' with hair like wool and a throne of fire. A mysterious 'Son of Man' who comes on the clouds of heaven to receive an everlasting kingdom. What does it mean?

As it turns out, more than people think—and in a way that is far more visually immediate than the technical theological debates around this chapter might suggest. Daniel was looking at the night sky. He was seeing the constellations that he and the Babylonian wise men he led had spent years cataloging and organizing, and he was using those familiar celestial images to describe something that the heavens have been proclaiming since creation: the great cyclical arc of human civilizations, from their rise in power to their fall in judgment.

Daniel's insight and methods of prophecy and visions were not always as esoteric as one might imagine. To be sure, Daniel was given special discernment from God on many occasions, but he also had a more basic recognition of the enduring revelation in the stars—the 'signs' of Genesis 1:14. Daniel's influence and insight extended far beyond his own time and the nebulous future. We can recognize his influence in the story of the birth of the Messiah. Some very basic questions to ask are: why would 'wise-men' from the east be the least bit concerned about the birth of a Jewish Messiah, and how would they even know? Daniel's authority over much of the goings-on of Babylon gave him the authority and ability to make those visions known to whomever he chose.

Though not specifically stated in the Book of Daniel, Daniel clearly knows of the coming Messiah—'one like the Son of Man,' 'the Prince of princes.' He was then able to relate that knowledge to the Babylonian priests of his day—specifically, the need to watch for the sign and the star of the coming Jewish Messiah. Matthew 2:1-2,7 records that the 'wise men' from the east (Zoroastrian/Persian Magi [G3097]) came to find Jesus because they had 'seen His star.' The Magi were Zoroastrian priests who were very familiar with Daniel's idea of a monotheistic God and receptive to his instruction—instructions they would hold close in their memory for over 500 years.

The unidentified proclaiming star may have been pretty much anything: a meteor or comet, the

Shekinah Glory of God that Moses saw in the burning bush (Exodus 3:1ff), the pillar of cloud and fire that led the Israelites in the desert (Exodus 13:21), or a supernatural manifestation like the glow of Moses' face after receiving the Commandments (Exodus 34:30), or the Transfiguration of Jesus Himself (Matthew 17:2).

The star did not lead the Magi to Jesus—it simply proclaimed His birth. In fact, the star in the east was opposite to the direction of Jesus's birthplace. As Daniel doubtlessly taught them, the star simply told them it was time to mount-up and get ready to ride toward Israel, see Matthew 2:2. Once in Jerusalem they had to inquire of His whereabouts. That prophecy, already known to the Pharisees, was Micah 5:2: 'But you, O Bethlehem of Ephrath, Least among the clans of Judah, From you shall come forth To rule Israel for Me, One whose origin is from of old, From ancient times.' It was not until they had searched the scriptures that the star reappeared over Jesus' exact whereabouts, Matthew 2:9-10.

The essence of Daniel's vision of cycles starts in chapter 7, where he begins to get very personal and his narrative changes to first person. Daniel opens chapter 7 saying (7:2), 'In my vision at night...'—so it is clear that what he sees is something visible at night. But it is also evident that, though he sees the images clearly, he does not fully understand their significance. Like John's visions in Revelation, there is someone else present to help with the complexity of the visions (Daniel 7:16 and Revelation 1:1).

One important caution before we proceed: do not confuse what follows with astrology and horoscopes. Daniel understood the patterns of stars as signs, as God intended them at creation. He uses the ancient, familiar constellations to refer to cycles of secular history—not to predict individual destinies or personality types. All of the animal images he sees are deliberately echoed in Revelation 13:2, confirming that these are not personal symbols but cosmic historical markers. Daniel's vision embodies a large portion of the more grand cycle of history — the kind of repeating pattern that the philosopher George Santayana had in mind when he observed that those who forget the past are condemned to repeat it. The cycle reads like this:

Daniel 7:2-3 — The Great Sea and the Four Winds

Daniel 7:2-3: 'four winds [H7308] of heaven [H8065] stirring up the Great Sea [H3221].' Though not specifically representing a single constellation, this opening image suggests that Daniel was looking into the depths of heaven where, in the language of creation itself, 'the Spirit of God was hovering over the face of the waters' (Genesis 1:2).

The Great Sea: Heaven or Mediterranean?

The image that opens Daniel's vision—four winds stirring up the Great Sea—deserves its own moment of consideration, because it is doing more work than it might first appear. The Hebrew term rendered 'Great Sea' is יַמָּא רַבָּא (yamma rabba) [H3221, H7229] in Aramaic—the form Daniel uses here, since the visionary sections of chapters 7 and 8 are written in Aramaic. The phrase could refer to the Mediterranean, which is how the Hebrew Bible typically uses 'Great Sea' in a geographical sense. But within the context of what follows—a vision of celestial creatures arising from cosmic depths—something larger is being evoked.

In biblical cosmology, the sea is not simply a body of water, it is the primordial deep—תְהוֹם (tehom) [H8415]—the formless, chaotic waters that preceded creation (Genesis 1:2). When the four winds of heaven stir this sea, Daniel is witnessing something that mirrors the very opening of creation: the Spirit moving over the face of the waters, and out of that formless deep, form arising. The beasts that emerge from the sea in Daniel's vision are therefore not creatures of the earthly Mediterranean. They are cosmic powers arising from the primordial depths of heaven.

This reading is supported by the parallel in Revelation 13:1, where John sees 'a beast rising up out of the sea, having seven heads and ten horns.' John's beast combines features of Daniel's beasts—leopard, bear, lion—in a single apocalyptic form. The sea from which it rises is the same cosmic deep. Both Daniel and John are looking into the same abyss: the vast, wind-churned expanse of heaven from which the great powers of history arise, and to which they ultimately return.

There is also a celestial geography to consider. In the night sky as Daniel would have seen it, the Milky Way—that diffuse band of starlight that crosses the sky—was sometimes described in ancient astronomical contexts as a great river or sea of heaven. The two points where the Milky Way intersects the ecliptic—the band within which Daniel's visionary constellations are arrayed—are identified in this book as the Clouds of Heaven (Daniel 7:13) and the River Ulai (Daniel 8:2). Between these two crossing points, the great constellations of Daniel's vision are arrayed like creatures rising from the depths of a celestial sea.

This is precisely why the beasts arise from the sea: they are the constellations arising from the depths of heaven, seen by a man who understood that the night sky was God's first and most permanent canvas for the telling of His story. The Great Sea is the vastness of heaven. The four winds are the cosmic forces that set history in motion. And the creatures that arise from its depths are the great empires—and their celestial counterparts—that mark the progress of the cycle from its beginning to its end.

There is one more detail worth noting. In Revelation, the sequence of Daniel's beasts is reversed. Where Daniel sees lion, bear, leopard, and bull in that order, Revelation 13:2 describes the apocalyptic beast as having 'the body of a leopard, the feet of a bear, and the mouth of a lion.' The order is reversed —and so is the direction. Daniel's vision moves forward through history, from Babylon toward Rome. Revelation's vision moves backward, from the apocalypse back toward the beginning. The beasts are the same beasts, read in reverse—a perfect illustration of the cyclical structure this book has been tracing from its opening pages. What Daniel saw coming, John saw going. The cycle turns.

7:4 - **Lion** [H744] representing the constellation Leo, is the Babylonian Empire (605 - 538 BC), but also a reference to the Lion of the Tribe of Judah (Revelation 5:5), the family of King David leading to the Messiah.

- The lion symbolizes the Babylonian Empire, reflecting its strength and majesty. The lion, king of beasts, mirrors Babylon's dominance under Nebuchadnezzar II (605–562 BC), whose conquests, including Judah (605, 597 & 586 BC), established a golden era (Daniel 2:38). Its "eagle's wings" suggest swift expansion, while their plucking indicates decline after Nebuchadnezzar, as weaker rulers like Belshazzar faltered (Daniel 5). The "heart of a man" hints at Babylon's humbling, as seen in Nebuchadnezzar's madness (Daniel 4:33).

7:5 - **Bear** [H1678] representing the constellation Ursa, is the Medo-Persian Empire (538 - 331 BC).

- The bear symbolizes the Medo-Persian Empire, embodying its massive, lumbering power. Raised on one side, it reflects Persia's dominance over Media after Cyrus the Great's rise (550 BC). The bear's three ribs in its mouth symbolize key conquests, Lydia (546 BC), Babylon (539 BC), and Egypt (525 BC), marking the empire's voracious expansion. Its slow, deliberate strength contrasts with Babylon's swift lion, as Medo-Persia's vast bureaucracy and military consolidated control from India to Ethiopia.

7:6 - **Leopard** [H5245] representing the constellation Lynx, is the Greek Empire (331 - 168 BC). Lynx, as a constellation is rather new. It was established by Johannes Hevelius, a Polish astronomer, in 1687. Due to the faintness of the stars in this location, previous to Hevelius, there were no constellations assigned to the space taken by Lynx.

- The leopard symbolizes the Greek Empire, capturing its agility and speed under Alexander the Great (336–323 BC). The leopard's four wings denote rapid conquests, as Alexander swept from Greece to India in a decade, forging a vast empire. Its four heads represent the division after his death, splitting into four kingdoms under his generals, those being: Cassander ruled over Macedonia/ Greece, Lysimachus ruled over Thrace (modern day Balkans), Seleucus ruled over the lands east of modern day Israel, and finally, Ptolemy ruled over Egypt and northeast Africa.

At this juncture a subtle but important shift occurs. While Daniel's visions continue unbroken, there is a break in the celestial narrative. Note the Milky Way — the hazy gray cloud that bisects the zodiacal band in two places in the northern sky. The first crossing is between Lynx and Taurus; the second is between Sagittarius and Scorpio. Both of these conjunctions have names that are hidden in their positions in the sky and their function within Daniel's visionary narrative. This is the Clouds of Heaven and the River Ulai. We will encounter both of them again.

7:7,8 - **Bull,** (beast [H2423]) implied as Taurus because "..it was different . . it had (ten) horns [H7162].", is the Roman Empire (168 BC - 476 AD). Interestingly, when shown correctly as a zodiacal image, only the front quarters of Taurus are shown. This is because the Roman Empire never really ended, like the other empires. Under Roman Emperor Constantine the Great, in 330 AD, the eastern Roman Empire became the Byzantine Empire, and continued the Roman Empire in the form of the Catholic Church.

- The fearsome beast with "horns" implied as a bull (Taurus), symbolizes the Roman Empire. Its "iron teeth" and trampling strength reflect Rome's unmatched military might, crushing nations from Britain to Egypt (168 BC–476 AD). The ten horns symbolize Rome's provinces or successor states, with the "little horn" possibly a later ruler or power (e.g., the Papacy).

7:9ff - **Lamb/Ram** (Aries), the sacrificial animal; (7:13) ".. one like the Son of Man", a foreshadowing of the Messiah. Daniel will be more specific about this image in chapter 8. Traditionally, Aries appears in the missing rear quarters of Taurus. This is because Jesus appeared in the time of Taurus and fundamentally transformed the Roman Empire (see the description of the Bull/Taurus above).

- Jesus is referred to as the Lamb in Revelation 13:8: "..names . . written in the Book of Life of the Lamb slain from the foundations of the world". Daniel, as continued in Revelation by John the Revelator, reveals the final imagery the brings the cycle back to its beginning.

The prophetic visions of Daniel 7 are not mere historical markers but celestial signposts, etched in the constellations as 'signs' of God's ordered plan. Strikingly, this imagery reappears in Revelation 13:2: 'The beast...was like a leopard, his feet were like the feet of a bear, and his mouth like the mouth of a lion.' This deliberate echo, penned centuries later, suggests more than continuity — it reveals a profound truth: the story from Genesis to Revelation is not a singular arc but a recurring cycle, unfolding across vast epochs, likely numerous times before.

Consider the parallels. Daniel's beasts, rooted in the 'four winds of heaven' stirring 'the Great Sea' (Daniel 7:2), mirror Revelation's chaotic sea-beast (Revelation 13:1), both culminating in divine judgment and renewal. The Lamb of Revelation 5:6—'a Lamb as though it had been slain'—echoes Daniel's 'Ancient of Days' (Daniel 7:9), tying redemption to creation's rhythm. This linguistic and symbolic unity suggests that Revelation is not a sequel to Daniel but a continuation—call it Daniel, Book 2. It retells the same saga of creation, fall, and restoration using the same celestial language and imagery, as if history turns on a cosmic wheel.

This cyclic view aligns with Genesis 1:1's 'when God began to create' and Revelation 21:1's 'new heaven and earth,' where the first passes away. Revelation extends this, projecting the cycle's end into a renewal, or perhaps another beginning. Daniel and Revelation are twin lenses—seeing the end and beginning simultaneously—revealing history's iterative cycles. Their shared imagery invites us to see the Bible as a unified story: not of one creation but of many, each restarting where the last concluded. In this light, Revelation is Daniel's echo, proclaiming that what was, is, and will be again, until God's eternal purpose finally shines forth (Isaiah 46:10).

3.3 Daniel 8: Bookends to the Christian Era

If Daniel 7 gave us the grand sweep—six celestial features accounted in order of their occurrence in the night sky, that account for the rise and fall of four empires from Babylon through Rome, culminating in the Messiah (the Lamb)—Daniel 8 zooms in. It focuses the lens on a specific segment of the cycle: the era that begins with the Lamb and ends with the apocalyptic goat. This is, in the most direct sense, the Christian era. And Daniel's description of it, written more than five centuries before Christ was born, is remarkable in its precision and its symbolic richness.

While the visions of Daniel 7 mark the periods of secular history, Daniel 8 serves as a captivating continuation of that imagery, weaving a visionary tapestry that extends the cyclic narrative to a new period of redemption. Daniel himself frames this vision as a continuation, noting it came 'after the one that appeared to me the first time' (Daniel 8:1).

Where Daniel 7 presented four beasts—lion, bear, leopard, and the horned beast—Daniel 8 narrows its focus to two figures that continue the zodiacal sequence: the Ram [H352]—the sacrificial Lamb identified with Aries—which follows in sequence after Taurus, and a he-goat [H5795/H6842] identified with Capricorn. This signals a deeper exploration of a specific era within the cosmic cycle, one that resonates as the Christian era beginnings with 'the Lamb who takes away the sin of the world' (John 1:29) and concluding in the apocalypse.

Daniel 8 opens with a specific and important note about his location (8:2). He is in the citadel of Shushan (Susa) by the river Ulai—known to the Greeks as the Eulaeus river, the Karun River of modern Iran. This setting is not merely a geographical detail, it is theologically resonant: Daniel stands on the banks of the very river that, within the cyclical celestial framework, represents the Milky Way's first crossing of the ecliptic—the passage between Lynx and Taurus/Aries. He is standing, symbolically, at the River Ulai of the heavens. The citadel's prominence, rising above Elam's fertile plains, reflects the imperial power that will soon be challenged. Rivers in Scripture often frame divine revelation (see Ezekiel 1:1), and here too, the setting announces that what follows is a vision of cosmic significance.

Within the cyclical paradigm, the River Ulai represents the junction of the Milky Way spreading as a wide nearly invisible river between Lynx and Taurus. Nestled within the flanks of Taurus, on the banks of the Ulai, is Aries—the sacrificial Lamb. We will see the Milky Way expressed again later as the Clouds of Heaven, from which redemption and final victory over evil descend.

In Daniel 8:4, the Ram is seen 'pushing westward, northward, and southward,' but notably not eastward, its horns growing 'very great.' This movement mirrors the expansion of the Christian church during the Piscean era—the age of 'fishers of men' (Matthew 4:19). From its origins in Judea, Christianity spread west to Europe, north to Asia Minor, and south to Africa, but its eastward push would eventually be thwarted by Islam. The Ram, as Aries, embodies Christ the Lamb, whose influence reshaped the Roman world and beyond. Yet paralleling this era is a shadow: the serpent-like Cetus, a sea monster lurking near Pisces' celestial sphere, symbolizing Satan's counterforce—Revelation 12:9: 'that serpent of old, called the Devil.'

The vision takes a darker turn in Daniel 8:5, as 'a he-goat came from the west...with a notable horn between his eyes.' This goat, identified with Capricorn, charges the Ram and shatters its horns (Daniel 8:7). The 'notable horn' grows 'very great' (Daniel 8:8), only to break, giving rise to four horns—echoing the Greek Empire's division after

Alexander the Great (323 BC), as in Daniel 7:6. But the goat's deeper significance lies in its apocalyptic role: it commits 'the transgression that makes desolate' (Daniel 8:13), halting Jewish and Christian worship in God's temple. This 'abomination of desolation' (Daniel 11:31) prefigures events like Antiochus IV's defilement of the Jerusalem temple in 167 BC or, in the cyclic view, later assaults on faith including early Roman persecutions and the ultimate rise of the Antichrist (Revelation 13:5-7).

Capricorn's dual nature complicates and enriches the imagery. In the traditional zodiac it is a chimaera —upper half goat, lower half dragon or fish— blending terrestrial and serpentine traits. The Hebrew text of Daniel 8:5 illuminates this duality through grammar. Unlike the English rendering 'male goat,' the Hebrew uses a maqaf—a hyphen-like connector—linking a 'she-goat' [H5795] and a 'he-goat' [H6842] into a single grammatical concept. This union, emphasizing the masculine 'he-goat,' conveys a single creature with dual character, mirroring Capricorn's hybrid form. The maqaf tends to stress the second element in the duality—in this case, the male goat imagery—underscoring the creature's role as a chimeric adversary blending deception and raw power to disrupt God's worship.

Daniel 8's core message is stark: the he-goat, as Capricorn, seeks to dismantle the Jewish and Christian worlds by abolishing sacred practices. Historically, this aligns with figures like Antiochus IV, who banned sacrifices and erected an altar to Zeus in the Jerusalem temple. Cyclically, it points to the tribulation's climax, where the 'little horn' (Daniel 8:9) magnifies itself against God's truth (Daniel 8:12). Daniel 8:14 sets a specific timeline: 'two thousand three hundred evenings and mornings; then the sanctuary shall be restored.' This span of roughly 6.3 years nearly covers the complete seven-year tribulation of Revelation 11:2-3, suggesting the goat's dominance consumes most of that final period.

So Daniel 8 is a continuation of Daniel 7 and a foreshadow of the events of Revelation. Within the cyclic framework, Daniel 8 maps the post-Roman zodiacal sequence: after the Piscean Christian era, Aquarius pours out judgment (Revelation 16), ushering in Capricorn's chimeric reign, before Sagittarius—Jesus on a white horse in the clouds of heaven (Revelation 19:11-16)—defeats the Adversary and restores God's order. Thus, Daniel 8 illustrates bookends to the Christian era with the Lamb's redemptive entrance and the goat's apocalyptic opposition, completing the cycle that brings us from the beginning of Aries to the end of

Capricorn, from the first coming of Christ to His victorious return.

3.4 Origins of the Zodiac and Connection to the Bible

The word 'zodiac' tends to land like a mild accusation in Christian circles. It carries the sense of horoscopes, newspaper astrology columns, and the New Age movement's appropriation of ancient symbols for purposes far removed from their origins. Before we examine how the zodiac connects to the Bible, it is worth being clear about what the zodiac actually is, where it came from, and why the biblical authors referenced it without embarrassment.

The zodiac, in its original form, was not about predicting individual futures, it was about mapping time. It was a calendrical and astronomical framework—the ancient world's most systematic attempt to track the movement of the sun, moon, stars and planets through the twelve positions that occupy the band of sky known as the ecliptic. Its origins are genuinely ancient, and they are deeply intertwined with the history of both Babylon and Egypt—the two great civilizations that frame Israel's own story.

Babylon: The First Systematic Astronomers

As we established in section 3.1, the Babylonians brought to the study of the stars a passion for organization that was characteristic of their civilization as a whole. The MUL.APIN tablets — the great star catalogs of the 7th century BC—represent the most systematic astronomical record produced anywhere in the ancient world up to that point. They listed the positions of stars, the rising and setting times of constellations through the year, the pathways of the moon and planets, and the relationships between celestial events and the agricultural calendar.

From this tradition emerged the formalization of the twelve-sign zodiac: twelve constellations arranged in sequence around the ecliptic, each occupying roughly 30 degrees of sky, each associated with a particular month of the year. The iconography of these twelve signs—the bull, the lion, the scorpion, the water-bearer, the fish—drew on an even older Sumerian tradition of stellar symbolism going back at least to 3000 BC. The Babylonians did not invent these images; they systematized and standardized them.

The resulting zodiacal system was a practical tool of enormous value. It allowed Babylonian astronomers to predict with reasonable accuracy the positions of celestial bodies months and years in advance. It provided a shared framework within which astronomical observations from different times and places could be compared and combined. And, it gave the ancient world—for the first time—a universal astronomical language that could be communicated across cultural and geographical boundaries.

Egypt: The Dendera Zodiac and a Parallel Tradition

While Babylon was developing its ecliptic-based zodiac, Egypt was developing its own tradition of stellar symbolism—one that shared remarkable similarities with the Babylonian system despite originating in a different culture and a different astronomical framework.

The most famous physical artifact of this Egyptian tradition is the Dendera Zodiac (see Image 3.1)—a circular bas-relief carving found on the ceiling of the Hathor temple at Dendera in Upper Egypt, dating from around 50 BC (though incorporating much older astronomical traditions). The Dendera Zodiac depicts all twelve signs of the familiar zodiac, along with numerous other constellations, arranged in a circular format centered on the north celestial pole. It is, in effect, a map of the ancient Egyptian sky.

What strikes the careful observer of the Dendera Zodiac is how familiar it looks. Despite originating in a completely different civilization, the twelve signs are instantly recognizable: the lion, the scorpion, the bull, the twins, the fish, the archer. The iconographic similarities between the Babylonian and Egyptian zodiacs are so pronounced that they cannot be explained purely by parallel independent development. There was contact, influence, and exchange — facilitated by the same trade networks and diplomatic relationships that moved grain, timber, lapis lazuli, and ideas around the ancient Near East for millennia.

Image 3.1 - Egyptian Dendera Zodiac
(photographically enhanced for clarity).
Louvre Museum, Paris, France

The most likely explanation for the similarity is not that one system copied the other directly, but that both drew on a much older, shared tradition of stellar symbolism that predated both civilizations—a tradition potentially associated with the proto-civilization that this book has been arguing existed before the previous cycle's apocalyptic conclusion. The relics of that earlier civilization—Göbekli Tepe, the Sphinx, the great megalithic sites of Europe and the Americas—suggest a level of astronomical sophistication that would have included, at minimum, a well-developed system of star-tracking and celestial symbolism. The Babylonian and Egyptian zodiacs may both be, in a sense, local adaptations of a much more ancient universal template.

The Progression of the Zodiac Through History

From its Babylonian origins, the zodiac was carried westward by the expanding Persian Empire and then by the conquests of Alexander the Great. Greek astronomers encountered the Babylonian zodiac and were profoundly influenced by it. They adopted the twelve-sign framework, adapted the iconography to their own mythological tradition, and applied their own mathematical sophistication to the task of refining its predictive accuracy.

The most consequential figure in this Greek development was Claudius Ptolemy, an Alexandrian Greek astronomer working in the 2nd century AD. His two great works—the Almagest (a comprehensive mathematical astronomy) and the Tetrabiblos (a treatise on the astrological interpretation of celestial phenomena)—became the definitive reference texts for both astronomy and astrology in the Western world for the next fourteen centuries. Ptolemy did not invent astrology, but he systematized it in a form that made it intellectually respectable—a discipline with mathematical foundations that could be taught, learned, and applied consistently.

It is here, in Ptolemy's Tetrabiblos, that the zodiac was definitively transformed from an astronomical framework for tracking time and celestial events into a personal predictive system for determining individual destinies based on the positions of planets at the time of birth. This transformation—from cosmic calendar to personal horoscope—is the development that the Bible consistently cautions against. Deuteronomy 4:19 warns Israel not to be 'lured into bowing down to them or serving them' when looking at the stars. Isaiah 47:13-14 mocks Babylon's 'astrologers, the star-gazers, the monthly prognosticators' who claim to predict personal fortunes. The problem was never with observing the stars. It was with treating the stars as the authors of human destiny rather than as signs of their Author's purposes.

Other Zodiacal Traditions Around the World

It is worth briefly acknowledging that the Babylonian-Egyptian-Greek zodiacal tradition is not the only star-based system of symbolic astronomy in the world. Indian astronomy developed the Nakshatras—a system of 27 or 28 lunar mansions that tracked the moon's progress through the sky in a manner parallel to but distinct from the solar zodiac. The Chinese developed a system of 28 Lunar Mansions alongside a twelve-year cycle of animals—the Chinese zodiac—that organized time in a completely different framework from the Western one. The indigenous peoples of Australia developed sophisticated star knowledge tied to their ecological seasons and navigational needs. The peoples of the Americas developed their own systems of stellar observation, many of which show levels of astronomical sophistication that continue to surprise modern researchers.

The diversity of these systems does not undermine the argument this book is making. It supports it. The universal human impulse to read the stars as meaningful — to look up at the night sky and see not random points of light but a language of divine purpose—is itself evidence that the stars were placed there to speak. Different cultures developed different vocabularies for reading that language, just as different cultures developed different natural languages for speaking to each other. But the impulse to look, to catalog, and to find meaning is universal. God made it that way. 'The heavens tell the glory of God, the sky proclaims His handiwork. Day to day makes utterance, night to night speaks out' (Psalm 19:2-3).

The Zodiac in the Hebrew Bible

When God fashioned the heavens, He imbued the stars with purpose, declaring, "Let there be lights [H3974] in the firmament of the heavens… for signs [H226] and seasons, and for days and years" (Genesis 1:14). If these celestial bodies were ordained as divine markers, we should find evidence of their role as such woven throughout scripture. Beyond Daniel's visions, which cast constellations as emblems of history's secular empires (Daniel 7:4–7), the Hebrew Bible reveals a surprising breadth of stellar awareness. From patriarchs to poets, Israel's writers gazed upward, discerning in the stars a narrative crafted by God, a story of creation, judgment, and redemption that transcends time and culture.

This celestial testimony aligns with God's eternal sovereignty, as proclaimed in Isaiah 46:10, "I foretell the end from the beginning, and from the start, things that had not occurred." If God is unchanging, His message should persist across the cycles of history, iterations of creation and renewal spanning millennia. How could such a timeless story endure, accessible to ".. all the peoples under the whole heaven as a heritage." (Deuteronomy 4:19)? The answer lies in the stars, God's immutable canvas. Unlike scrolls that fade or stones that erode or crumble, the heavens proclaim His plan nightly, their patterns are a universal language for every generation, in every cycle.

The Hebrew Bible abounds with evidence of this stellar awareness. Job, perhaps the earliest Biblical story, marvels at God's handiwork: "He made the Bear [H5906], Orion [H3685], and the Pleiades [H3598], and the chambers of the south" (Job 9:9). Later, he asks, "Can you bring out Mazzaroth [H4216] in its season? Or can you guide the Great Bear with its cubs?" (Job 38:32). This awareness predates the Hebrews as a people, and Israel as a nation.

The Mazzaroth—broadly speaking, the Hebrew zodiac—suggests a recognized system of constellations, not for pagan divination but for divine storytelling. Amos 5:8 echoes this: “He made the Pleiades [H3598] and Orion [H3685]”, tying stars to God’s creative power. Even the “host of heaven” [H6635, H8064] (2 Kings 23:5) reflects a cosmic order, though Israel was warned against worshiping it (Deuteronomy 4:19).

The following are direct references to the zodiac in the Hebrew Bible (JPS):

Mazzaroth [H4216] - broadly, the zodiac as referenced in the Hebrew Bible.

- Job 38:32: ”Can you lead out Mazzaroth in its season? Or conduct the Bear (ayish) with her sons?”

Mazzaloth [H4208] - constellations

- 2 Kings 23:5: ”.. to the sun, to the moon, to the constellations (mazzaloth), and to all the host of heaven.”

Kesil [H3685] - Orion

- Job 9:9: "Who made the Bear (ayish) and Orion (kesil), Pleiades, (kimah) and the chambers of the south wind."

- Job 38:31: ”Can you tie cords to Pleiades (kimah), or undo the reins of Orion (kesil)?”

- Amos 5:8: "Who made the Pleiades (kimah) and Orion (kesil).."

- Isaiah 13:10: "For the stars of heaven and their constellations (kesilim, *plural*) will not give their light..."

Kimah [H3598] - the Pleiades

  - Job 9:9: "Who made the Bear (ayish) and Orion (kesil), Pleiades, (kimah) and the chambers of the south wind."

  - Job 38:31: "Can you tie cords to Pleiades (kimah), or undo the reins of Orion (kesil)?"

  - Amos 5:8: "He made the Pleiades (kimah) and Orion (kesil).."

- **Ayish** [H5906] - the Bear

  - Job 9:9: "Who made the Bear (ayish) and Orion (kesil), Pleiades, (kimah) and the chambers of the south wind."

  - Job 38:32: "Can you lead out Mazzaroth in its season?
    Or conduct the Bear (ayish) with her sons?"

The following are possible indirect references to the zodiac in the Hebrew Bible (JPS):

Host of Heaven - tseva hashamayim [H6635 H8064]

- Deuteronomy 4:19: "And when you look up to the sky and behold the sun and the moon and the stars, the whole heavenly host (tseva hashamayim), you must not be lured into bowing down to them or serving them.."

- 2 Kings 23:5: ".. to the sun, to the moon, to the constellations - all the host of heaven (tseva hashamayim)."

- Also, Jeremiah 8:2, Zephaniah 1:5

Chambers of the South - chadrei teman [H2315 H8486]

- Job 9:9: "Who made the Bear (ayish) and Orion (kesil), Pleiades, (kimah) and the chambers of the south wind (chadrei teman)"

Rahab - rahab [H7294], as a serpent or dragon, Draco

- Job 26:12: “By His power He stilled the sea; By His skill he struck down Rahab (rahab)."

- Psalm 89:10: You crushed Rahab; he was like a corpse; with Your powerful arm You scattered Your enemies.

- Isaiah 51:9: “.. It was you that hacked Rahab in pieces, that pierced the Dragon?

Elusive Serpent - nachash bariach [H5175 H1281] - serpent

- Job 26:13: By His wind the heavens were calmed;
  His hand pierced the Elusive Serpent (nachash bariach).”

3.5 The Complete Cycle from Genesis to Revelation

The prophet Daniel's visions in Daniel 7 and 8 reveal a profound truth: the Hebrew Bible engages the starry heavens with a freedom that defies the rigid ecliptic zodiac codified by later Babylonian astronomers. Daniel envisions the constellations not only as emblems of empires (Babylon, Medo-Persia, Greece, Rome) but as a divine narrative written for everyone to see. Daniel's creative liberty underscores a deeper reality: the stars, ordained by God as "signs and seasons" (Genesis 1:14), tell an enduring story of creation, fall, and redemption, a saga etched in the heavens long before Mesopotamian astronomers mapped the ecliptic zodiac.

This celestial narrative transcends cultural boundaries and historical epochs. The Hebrew Bible confirms this awareness: Job 9:9 & 38:32 recognizes the "Mazzaroth" and "the Bear", while Amos 5:8 names "Pleiades and Orion". These are not pagan musings but acknowledgments of God's cosmic design, a narrative that persists across the cycles each marked by creation's dawn and renewal's close.

You may also notice, not all of the zodiacal images referenced in Daniel and Revelation are directly on the ecliptic plane. This too, is an important clue. There are innumerable other constellations and images in the night sky. In other cycles these constellations may play a different roll in illustrating the history of that cycle. In other cycles Perseus, Hercules, Orion, Andromeda, Pegasus; as well as, Cancer and Gemini may have played a defining roll in the narrative of another version of Daniel and John.

This book seeks a more consistent Biblical view focused on the current cycle, anchoring zodiacal images to Biblical themes—creation in Virgo, redemption in Aries, justice in Libra—while respecting traditional associations, like Daniel's beasts as ancient empires. These fluid meanings do not dilute the stars' divine purpose, they amplify it, reflecting God's sovereignty across cultures and cycles. Each cycle resets history, leaving relics — Göbekli Tepe's pillars, the Sphinx, the sarsens of Stonehenge, the Moai of Easter Island, Machu Picchu, Teotihuacan, Cusco, and countless others— as silent witnesses.

By anchoring Virgo to Genesis 1:1's "beginning" and Libra to Revelation's final justice, this book unveils a unified Biblical narrative. The stars, unlike empires or monuments, endure across cycles, their patterns a divine heritage offered to “all the peoples” (Deuteronomy 4:19) inviting us to read the heavens through scripture’s lens, not cultural or personal whims. As Psalm 19:1 proclaims, “The heavens declare the glory of God,” and narrates a story that loops from Revelation back to Genesis, each constellation a chapter in the cosmic chronicle of God’s unchanging purpose.

As we progress through the images keep in mind, the zodiac is intended to tell a story not be a time keeper. The 2160 year period assigned to each constellation is irrelevant. The important matter is what aspect of the overall story the constellation represents.

It is also worth noting, to anticipate some confusion, the constellations progress through the night sky in a clockwise sequence. You will eventually notice that the story they tell—as I am presenting here—progress in a counterclockwise sequence. The counterclockwise sequence is how the sun progresses through the zodiacal positions over the long 25,920 year (2160 year x 12 zodiacal images) Precession of the Equinox.

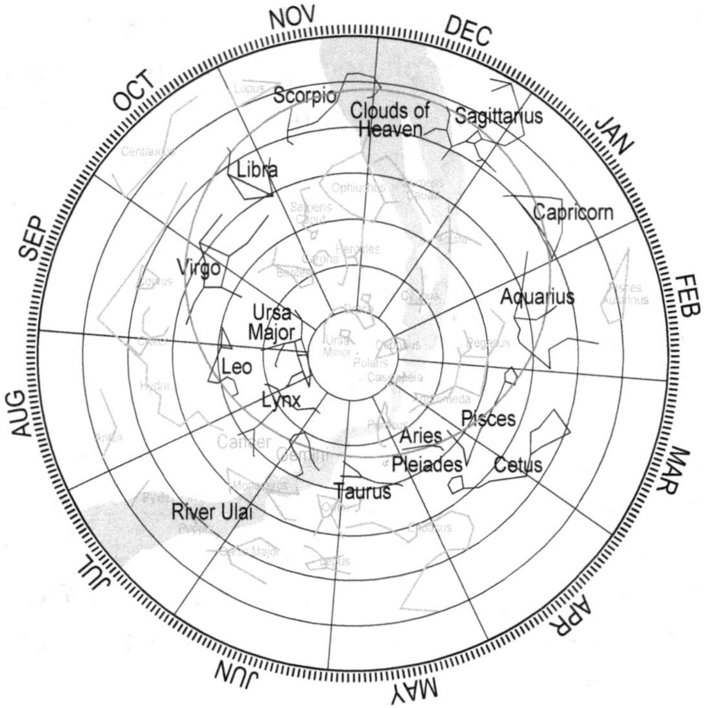

Figure 3.1 - Constellations of the Northern Hemisphere.

What follows is the complete story of Biblical cycles, as told in the stars (Genesis 1:14):

If the constellation images appear at odd angles that is simply an adjustment to allow them to fit within the constraints of the page. Nothing is implied by the changes in orientation with respect to their positions in the night sky.

Virgo - the new beginning (Genesis 1 & Revelation 21), creation in its perfect form. Later, the purity of the Redeemer's birth. The woman/virgin appears in Revelation 12:14, "..the woman was given two wings of a great eagle..", and as described progressively throughout Revelation 12. She represents God's redemption overcoming the efforts of the beast to thwart God's plan.

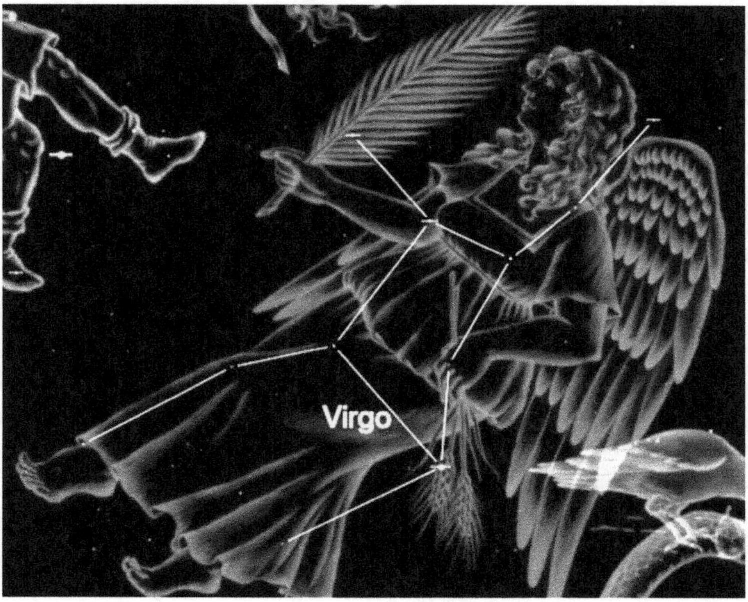

Figure 3.2 - Virgo.
Image courtesy Starry Night Astronomy.

Historically - Virgo is one of the zodiac's largest constellations and traces its history to ancient Mesopotamia, where it was associated with fertility goddesses like Ishtar. In Greek mythology it is linked to Astraea, goddess of justice. Its bright star Spica—one of the brightest in the night sky—made it easily identifiable across cultures and millennia. Ptolemy cataloged it in the 2nd century AD. That Virgo both opens the creation cycle and represents the virgin birth of the Messiah is one of the most elegant convergences in this entire celestial narrative.

While Virgo represent purity and a new beginning, what follows is a sequence of constellations that represent history, as defined by Daniel.

Leo / Daniel 7:4 - **Lion** [H744] - represents both the Lion of the Tribe of Judah (Revelation 5:5)—the family of King David and the lineage of the Messiah —and, in Daniel's vision, the Babylonian Empire (605-538 BC). This image appears in Daniel 7:4 and Revelation 13:2. The lion, king of beasts, mirrors Babylon's dominance under Nebuchadnezzar. Its 'eagle's wings' suggest swift expansion; their plucking indicates decline; the 'heart of a man' hints at humbling, as seen in Nebuchadnezzar's madness (Daniel 4:33).

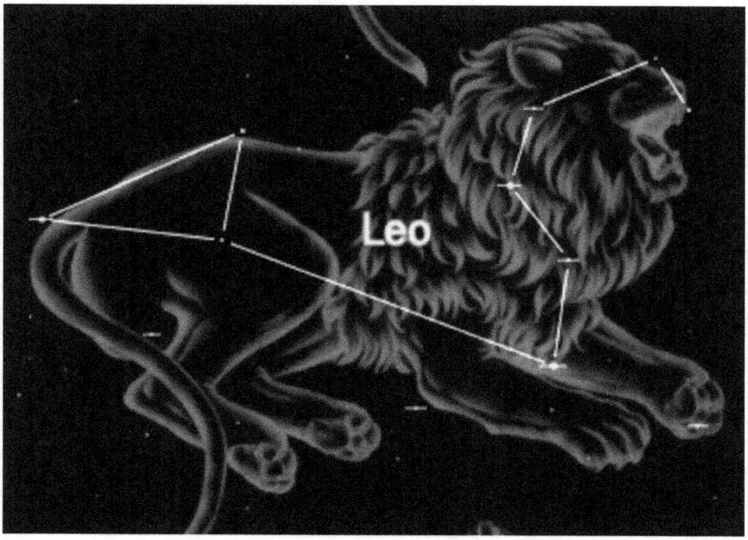

Figure 3.3 - Leo.
Image courtesy Starry Night Astronomy.

Historically - Leo is one of the oldest recognized constellations, associated with the great royal beast across virtually every ancient culture. In Mesopotamia, the lion was the symbol of the sun and of royal power. In Egypt, it was associated with the flood season when the sun was in Leo. The Sphinx—if scholars like Robert Schoch and John Anthony West are correct about its astronomical orientation—may have faced Leo in an earlier cycle. The constellation's prominent sickle of stars forming the lion's head and mane made it unmistakable in the ancient night sky.

Ursa / Daniel 7:5 - **Bear** [H1678] - (the Great Bear, or Ursa Major) represents the Medo-Persian Empire in Daniel 7:5, and appears again in Revelation 13:2. Raised on one side, it reflects Persia's dominance over Media. The three ribs in its mouth mark the three great conquests: Lydia, Babylon, and Egypt. Its slow, lumbering power contrasts with Babylon's swift lion.

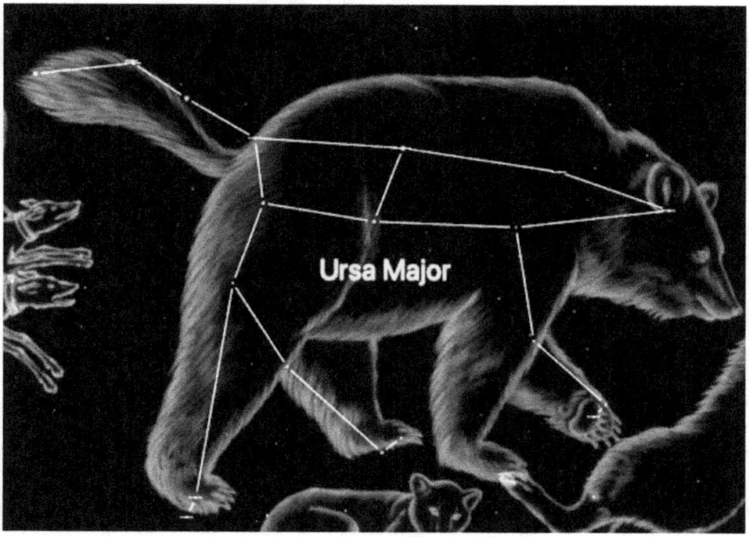

Figure 3.4 - Ursa.
Image courtesy Starry Night Astronomy.

Historically - Ursa Major is among the oldest recognized constellations in the world, documented across prehistoric cultures from Europe to North America to East Asia. Its seven brightest stars form the asterism known as the Big Dipper or the Plough, making it one of the most easily recognized patterns in the northern sky. In antiquity, the Great Bear's circumpolar position—it never sets below the horizon at mid-northern latitudes—made it a reliable navigation aid and a symbol of constancy. For the Hebrews, the Bear (Ayish [H5906]) appears in the oldest book of the Bible, Job 9:9, alongside Orion and the Pleiades, confirming that this constellation was part of Israel's earliest stellar vocabulary.

Leopard/Lynx / Daniel 7:6 - **Leopard** [H5245] - the constellation Lynx, established formally by Johannes Hevelius in 1687, fills a region of sky where no dominant stars form obvious patterns. Daniel's vision identifies this space with a leopard—the Greek Empire (331-168 BC), swift, agile, and multi-headed. The four heads of Daniel's leopard correspond to the four kingdoms that emerged from Alexander's empire after his death. Those kingdoms emerged from the rule of his four generals: Cassander ruled over Macedonia/Greece, Lysimachus ruled over Thrace (modern day Balkans), Seleucus ruled over the lands east of modern day Israel, and finally, Ptolemy ruled over Egypt and northeast Africa. The 'four wings' speak to the breathtaking speed of Greek conquest.

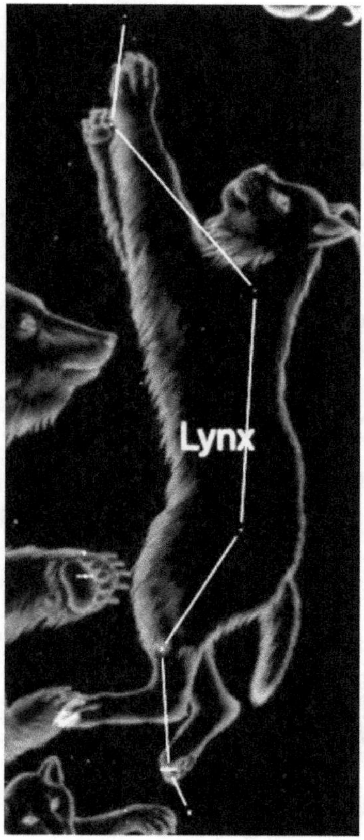

Figure 3.5 - Lynx.
Image courtesy
Starry Night Astronomy.

Historically - the Lynx region's faintness is precisely why Hevelius chose the name he did—he joked that only someone with the eyes of a lynx could see the stars there. Daniel's visionary eye saw something in that empty space millennia before any modern astronomer assigned a name to it. Much more than coincidence, Lynx is a striking example of Daniel's prophetic vision synchronizing with modern science.

Officially recognized by the International Astronomical Union in 1922, Lynx is one of the 88 modern constellations. Its dim stars, barely visible without a telescope, reflect the stealthy nature of its namesake. Notable features include the globular cluster NGC 2419 and the distant Lynx Arc, a massive star-forming region. Lynx remains an obscure yet intriguing constellation, observable primarily under dark skies, contributing to its mystique in astronomical history.

The River Ulai - The Milky Way:

At this juncture a subtle but important shift occurs. While Daniel's visions continue unbroken there is a break in the celestial narrative. Refer to <u>Figure 3.1 - Constellations of the Northern Hemisphere</u>. Note the hazy gray cloud that bisects the zodiac in two places.

As referred to in Daniel 8:2, barely visible between Lynx and Taurus is the first passage of the Milky Way. This is the first of two conjunctions where the Milky Way crosses the celestial zodiacal plane. The second is the **Clouds of Heaven** (Daniel 7:13, Matthew 24:30 & 26:64, Mark 14:62), where we look to see the return of the Savior.

It is the boundary between the era of the great empires (Daniel 7's sequence) and the era of the Messiah (Daniel 8's sequence). In chapter 8, standing on its banks, Daniel receives his vision of what the Christian era will look like.

History is about to focus much more sharply...

Taurus / Daniel 7:7,8 - **Bull,** (beast [H2423]) implied as Taurus because "..it was different . . it had (ten) horns [H7162].", is the Roman Empire (168 BC - 476 AD). Interestingly, when shown correctly as a zodiacal image, only the front quarters of Taurus are shown. This is because the Roman Empire never

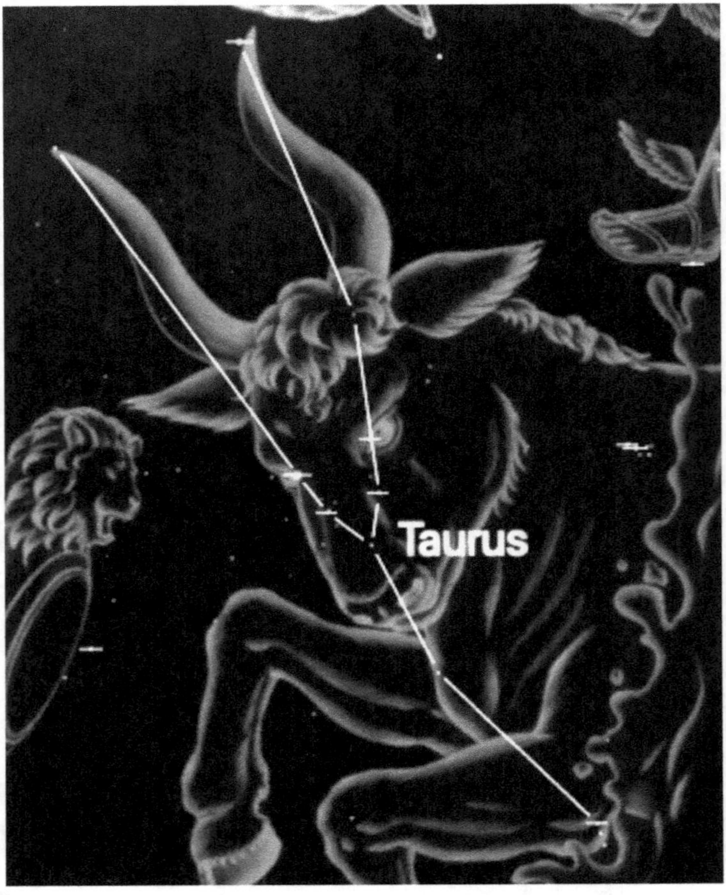

Figure 3.6 - Taurus.
Image courtesy Starry Night Astronomy.

really ended, like the other empires. Under Roman Emperor Constantine the Great, in 330 AD, the eastern Roman Empire became the Byzantine Empire, and continued the Roman Empire in the form of the Catholic Church. Its “iron teeth” and trampling strength reflect Rome’s unmatched military might, crushing nations from Britain to Egypt (168 BC–476 AD). The ten horns symbolize Rome’s provinces or successor states, with the “little horn” possibly a later ruler or power (e.g., the Papacy).

Historically - Taurus is one of the oldest documented constellations, identifiable in cave paintings at Lascaux, France, dating to approximately 15,000 BC. The Pleiades cluster, which sits on Taurus's shoulder, are mentioned in the Bible (Job 9:9, 38:31; Amos 5:8) and were important agricultural markers across cultures worldwide. That the very constellation associated with the Roman Empire appears to be among the oldest recognizable star patterns in human visual history is itself a form of ancient testimony.

The life and ministry of Jesus occur in the time of the Roman Empire, but His followers are encourage to continue His earthly ministry and mission of redemption!

Aries / Daniel 8:3ff - **Lamb** [H7716] or **Ram** [H352], the sacrificial animal that appears within the flanks of Taurus. As told in Genesis 22:1-14, Abraham was asked to sacrifice Isaac as a show of devotion. In the process of enacting that sacrifice both Abraham and Isaac are asking about a sacrificial **Lamb**, but in the end, God provided a **Ram** so Abraham did not have to sacrifice Isaac (Genesis 22:13). This story is symbolic of God offering His son Jesus Christ, the Messiah, the sacrifice to take away the sins of those who believe in Him. As He appeared near the end of the traditional military Roman Empire and the

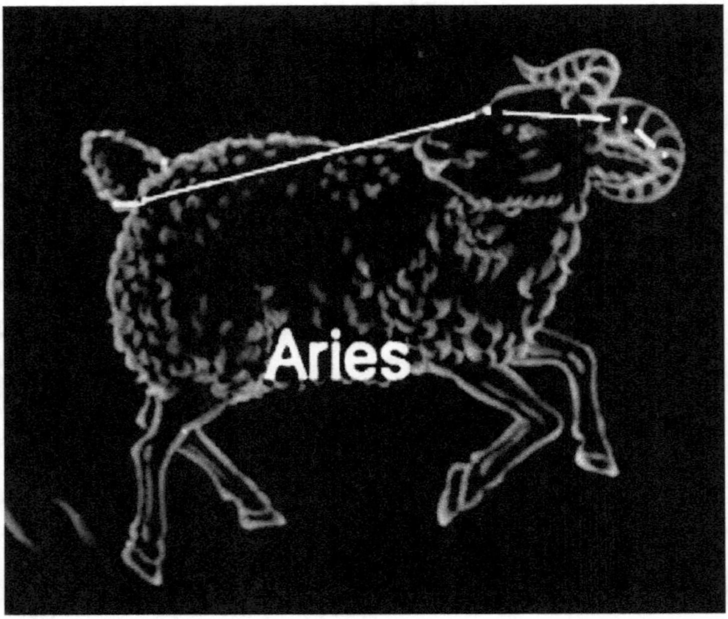

Figure 3.7 - Aries.
Image courtesy Starry Night Astronomy.

beginning of the Holy Roman Empire, this image appears in Daniel 8:3ff and throughout the book of Revelation. Aries appears in the missing hind quarters of Taurus. See Taurus for and explanation and significance.

Jesus is referred to as the Lamb in Revelation 13:8, "..names . . written in the Book of Life of the Lamb slain from the foundations of the world". Further, throughout Revelation Jesus is depicted as the Lamb that is slain, an avenger of blood, the object of worship, Lord of lords/King of kings, the bridegroom, the author of life, the source of light, the victor over death and separation from God! He is, "..the Alpha and the Omega, the Beginning and the End, the First and the Last." He begins the cycles and He ends the cycles.

Historically - Aries was the first sign of the zodiac in the ancient world when the spring equinox occurred in that constellation—a position it held from approximately 2000 BC to 0 AD, the very era of Israel's covenant history and the birth of Christ. This is not coincidental in the cyclical framework. The spring equinox in Aries—the beginning of the new year marked by the Lamb constellation—corresponds precisely to the era in which the Lamb of God appeared in human history. God had been marking the appointment in the stars for two thousand years before the event arrived.

In Daniel 8:4, the Ram is seen “pushing westward, northward, and southward”, but notably not eastward, its horns growing “very great.” This movement mirrors the expansion of the Christian church during the Piscean era, the age of “fishers of men” (Matthew 4:19). From its origins in Judea, Christianity spread west to Europe, north to Asia Minor, and south to Africa, but its eastward push would eventually be thwarted by Islam.

In Acts 2, following the death and resurrection of Jesus, the Holy Spirit appears like a wind, and the appearance of tongues comes to echo the healing of the nations still set for the future. All those present hear this new Godspell, the Gospel, in their own language. A new era has just been given a mission and a voice!

Pisces - represents the age of the 'fishers of men' (Matthew 4:19)—the Christian era as it unfolded through the church's mission in the world to this day. The fish, tied together by a cord, speak to the unity of believers across time and geography, all caught by the same Gospel, all bound to the same Messiah.

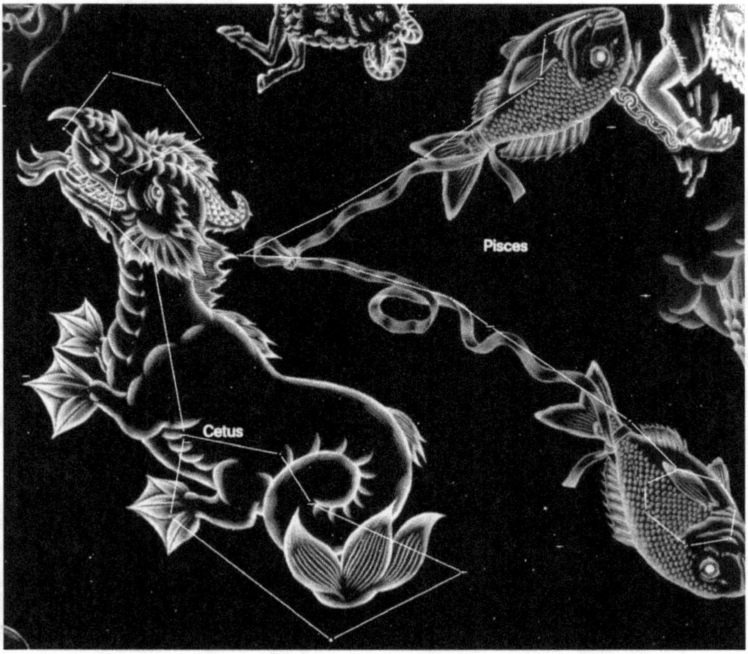

Figure 3.8 - Pisces/Cetus.
Image courtesy Starry Night Astronomy.

Cetus - Running parallel to Pisces in the sky is—the sea monster, the great whale or dragon of the southern sky. In the cyclical framework, Cetus represents the Adversary during the Christian era: Satan's counterforce to the Gospel, the serpent of old whose activity throughout the church age is described in Revelation 12:9—"..the great dragon was cast out, that serpent of old, called the Devil and Satan..." The Christian era has never been a simple story of uncontested advance. For every spread of the Gospel, there has been corresponding opposition. Pisces and Cetus exist in the same region of sky for the same reason.

Historically - Pisces traces back to ancient Mesopotamia, symbolizing fish tied by a cord. In Greek mythology it represents Aphrodite and Eros escaping Typhon. Cetus, the sea monster of Greek mythology, was the creature sent to devour Andromeda—a cosmic adversary defeated by the hero Perseus. The mythological resonance with the biblical drama of the Christian era is striking, even if the Greek myths have inverted and obscured the original meaning.

In Matthew 24 Jesus promises that all this will come to an end when He returns. His return will have its signs and precursors. A final accounting and the judgement of God, as told in the book of Revelation, are inevitable.

Aquarius - contrary to popular belief, this image does not represent the dawning of an age of enlightenment, but rather the pouring out of judgement. The water-bearer pours out not human wisdom but divine justice—the beginning of the apocalyptic judgment described in Revelation 16, where the angels pour out their bowls of God's anger upon the earth. Ezekiel 20:13 & 21 describe God's judgment on Israel for desecrating His

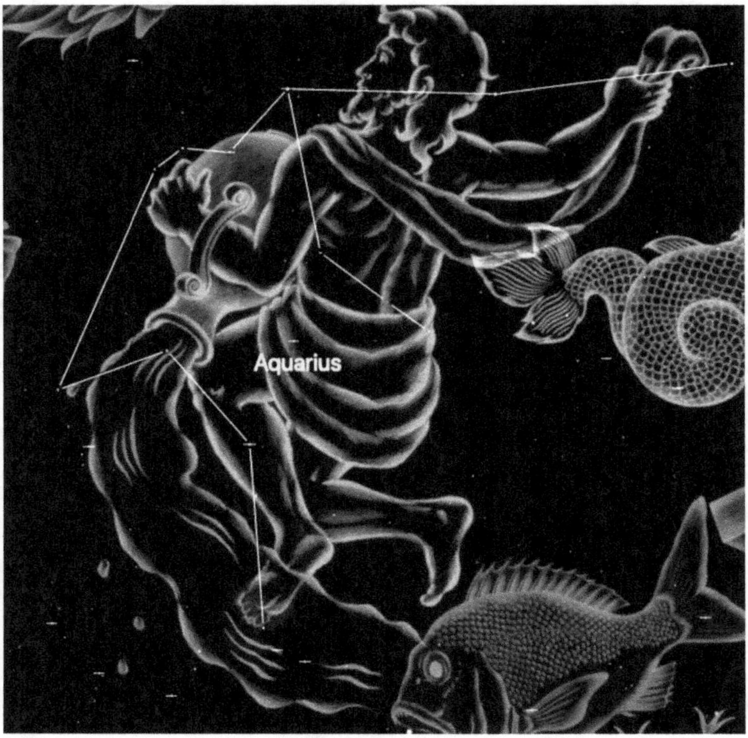

Figure 3.9 - Aquarius.
Image courtesy Starry Night Astronomy.

sabbaths—an appropriate echo, since it was precisely the missed sabbath rests that brought the Babylonian exile in the first place.

Historically - Aquarius traces its origins to ancient Mesopotamia, symbolizing a water-bearer linked to floods and renewal. In Greek mythology it represents Ganymede, Zeus's cupbearer. The image of a figure pouring water from a jar—whether the water is blessing or judgment—is one of the most ancient and universal symbols in human religious art. In the cyclical framework, the pouring begins where the peace of the Piscean era ends, and the great reckoning begins.

Some say we are currently in the age of Aquarius, citing the date of 2012 AD. This corresponds to the Mayan calendar reset date. Other say it won't happen until around 2600 AD. There does not seem to be any real consensus.

Conflict, contradiction and confusion will become the order of the day as evil makes its last stand, making it fairly clear that we are already in Aquarian era.

Capricorn - half goat, half dragon or fish—Capricorn is a chimaera, a hybrid creature that embodies the duality of the antichrist: outwardly religious and respectable (the goat), inwardly serpentine and destructive (the dragon/fish). This represents the beast of Revelation 13, the 'little horn' of Daniel 8:9 that magnifies itself against God and seeks to abolish His worship. In the Hebrew text of Daniel 8:5, this creature is described using two words connected by a maqaf, a word connector that links the two words into a single concept. Those two images are a 'she-goat' [H5795] and 'he-

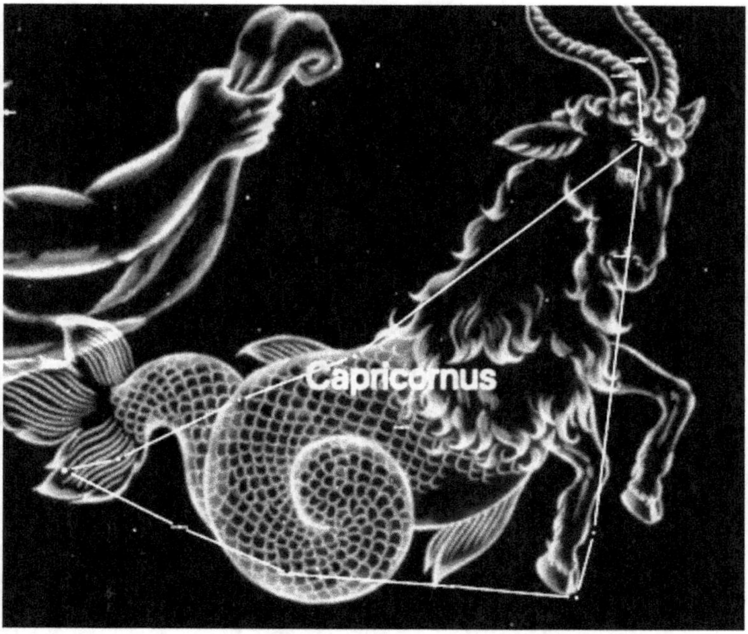

Figure 3.10 - Capricorn.
Image courtesy Starry Night Astronomy.

goat' [H6842]. They are combined into a single compound concept that emphasizes the dual nature of this entity: a chimeric adversary that cannot be simply categorized.

The goat of Daniel 8:5, identified with Capricorn, charges the Ram, shattering its horns (Daniel 8:7). The "notable horn" grows "very great" (Daniel 8:8), only to break, giving rise to four horns, echoing the Greek Empire's division after Alexander the Great (323 BC), as in Daniel 7:6. But the goat's deeper significance lies in its apocalyptic role: it commits "the transgression that makes desolate" (Daniel 8:13), halting Jewish and Christian worship in God's temple. This "abomination of desolation" (Daniel 11:31) prefigures events like Antiochus IV's defilement of the Jerusalem temple in 167 BC or, in a cyclic view, later assaults on faith, such as early Roman persecutions or the rise of the Antichrist (Revelation 13:5–7).

Historically - Capricorn traces its roots to ancient Mesopotamia, symbolizing a sea-goat associated with the god Ea, linked to wisdom and waters. In Greek mythology it represents the goat Amalthea, who nursed Zeus, or Pan transforming to escape Typhon. The image of a creature that is simultaneously of the earth and of the sea—capable of moving in both realms, belonging fully to neither

—is a perfect symbol for the shape-shifting nature of the antichrist who appears righteous while serving destruction.

All is not lost! There is a hero on a white horse coming to the rescue. Maybe the image is a bit confused, but there is method to the madness.

Sagittarius - half horse, half man, the Centaur (in Greek, kentaur)—represents Jesus Christ mounted on a white horse (Revelation 19:11-16, NKJV) coming in the clouds of heaven to vanquish evil. The Centaur's arrow flies across the Clouds of Heaven—the second crossing of the Milky Way—aimed directly at Scorpio, the source of torment and pain. In Greek and Roman mythology, the righteous

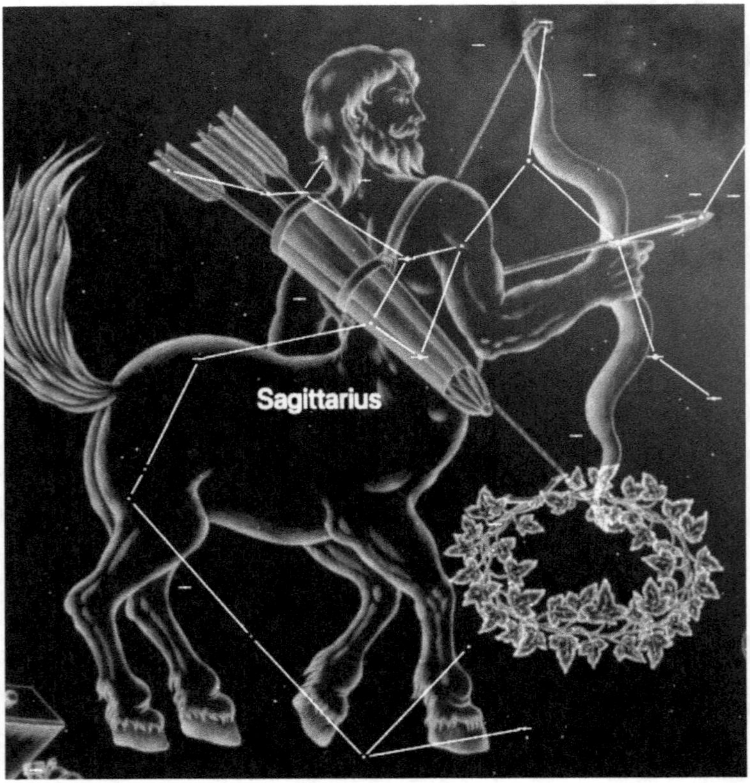

Figure 3.11 - Sagittarius.
Image courtesy Starry Night Astronomy.

centaurs were associated with wisdom, healing, and the protection of the innocent.

In Robert Bowie Johnson Jr.'s book The Parthenon Code: Mankind's History in Marble, Mr. Johnson interprets ancient Greek/Roman mythological characters as the inverse of many corresponding Biblical characters. In his interpretation of the centaurs Mr. Johnson says, "Before the Flood, the line of Seth intermarried with the line of Kain, and over time merged into it, leaving Noah's family as the only one with an untainted connection to Seth. Those events are related in Chapter 6 of the Book of Genesis. The Greeks remember these same events as the stealing of the line of Kain's women by Kentaurs, and the they depicted this embarrassment on the south metopes of the Parthenon."

Historically - Sagittarius is among the most ancient recognized constellations, associated in Mesopotamia with the god Nergal and later with the archer-figure across multiple cultures. In the night sky, Sagittarius aims his arrow across the center of the Milky Way toward Scorpio. That the constellation, representing Christ's victorious return, aims his arrow toward the galactic center—and across the Clouds of Heaven—is the kind of celestial detail that makes one pause.

The Clouds of Heaven - The Milky Way - the second conjunction where the Milky Way crosses the ecliptic plane falls between Sagittarius and Scorpio—and in Daniel's vision, this is 'the Clouds of Heaven' from which the Son of Man comes (Daniel 7:13; Matthew 24:30; Matthew 26:64; Mark 14:62; Revelation 1:7). Jesus has promised to return in exactly this way, and the celestial geography of the night sky places that return at precisely the point where Sagittarius—the archer on a white horse—crosses from the visible zodiac into the cloudy haze of the Milky Way. The promise and the sign are written together in the same region of sky.

Historically - The Milky Way, our spiral galaxy, is thought to have formed ~13.7 billion years ago. Ancient cultures, like the Greeks, named it "milky circle." Its barred spiral structure was confirmed in the 20th century. The Hebrews in Babylon (Daniel 7:10, from Aramaic) called it “the Fiery River”, in India it is called “the Heavenly Ganges”, in China “the Silver River”, the Andeans “the Celestial River”; as a path or road to the Lakota “the Spirit Road”, to the Cheyenne “the Hanging Road”, Finland “the path of the Birds”, Spain “the Road to Santiago”, Norse “the road to Valhalla”.

The end is now very near!

Scorpio / Scorpions (or scourges)—represents divine judgment in all its severity. The Hebrew aqrav [H6137] appears in passages associated with danger, punishment, and the harsh realities of the fallen world: Deuteronomy 8:15, 1 Kings 12:11 and 14, Ezekiel 2:6, and Revelation 9:3ff. In the cyclical narrative, Scorpio is the apex of the apocalyptic judgment: the full consequence of human rebellion against God, concentrated in one terrible image. Sagittarius aims his arrow here; the rider on the white horse comes to defeat this darkness and restore peace.

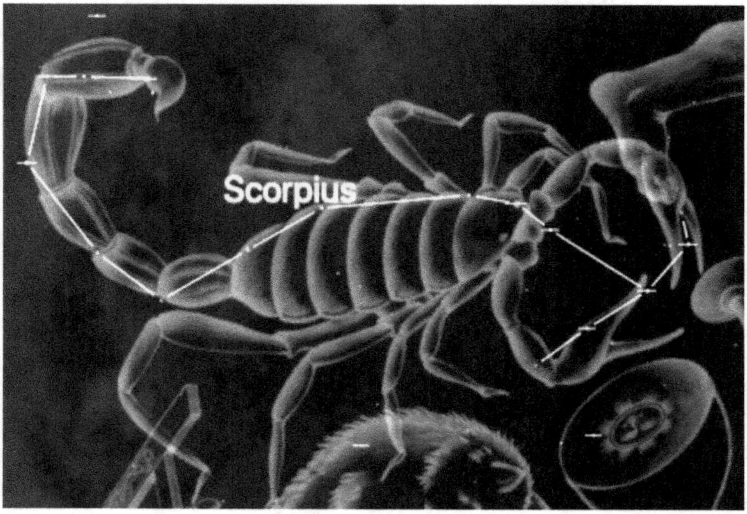

Figure 3.12 - Scorpio.
Image courtesy of Starry Night Astronomy.

Historically - Scorpio is one of the most ancient and consistently recognized constellations, clearly identifiable from its curved tail of bright stars. It originates in ancient Mesopotamia, where it was associated with death and transformation. In Greek mythology, Scorpio is the creature that killed Orion—placed by the gods in the sky opposite him, so that when one rises the other sets. Its association with transformation, danger, and the forces of darkness makes it a fitting symbol for the final, concentrated evil of the apocalyptic tribulation.

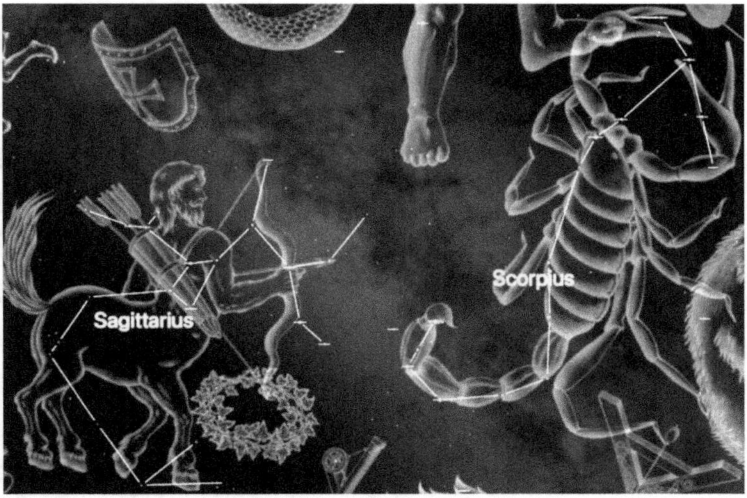

Figure 3.13 - Sagittarius/Milky Way/Scorpio cluster.
Image courtesy Starry Night Astronomy.

Sagittarius is aiming his arrow, shooting across - the **Clouds of Heaven** (the Milky Way) - to defeat the stinging evil represented by **Scorpio**. This is the essence of the Apocalypse of Revelation.

Libra / scales or balance - moznay [H3976], also peles [H6425]—the scales of justice, represents both the final victory of God's perfected plan and, simultaneously, the new beginning as the cycle turns back to Virgo. This image appears most powerfully in Revelation 6:5, where the third horseman carries scales. The justice of God that Libra represents is both terminal and generative: it concludes one cycle and initiates the next. For 'scales' of justice in the Hebrew Bible, see Leviticus 19:35-36, Job 31:6, Proverbs 11:1 and 16:11, Isaiah 40:12, and most memorably in Daniel 5: the

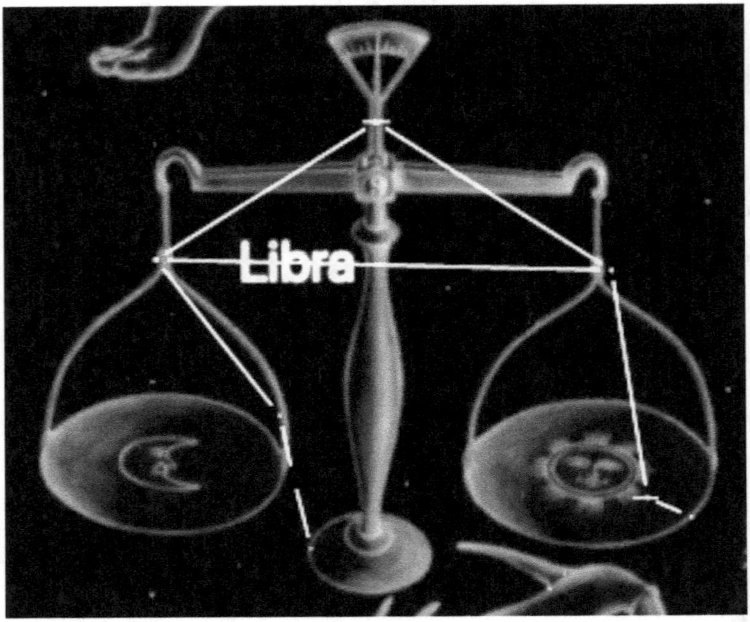

Figure 3.14 - Libra.
Image courtesy Starry Night Astronomy.

handwriting on the wall—tekel [H8625], meaning 'weighed in the balance'—is the scales of Libra applied to a king and a kingdom found wanting.

Historically - Libra traces its origins to ancient Mesopotamia, symbolizing balance scales, and in Greek mythology is linked to Astraea, goddess of justice. It was the only zodiacal sign representing an inanimate object rather than a creature—as if the ancients understood that at this point in the cycle, it is not human power or animal instinct that matters, but the cold, precise, and impartial justice of God.

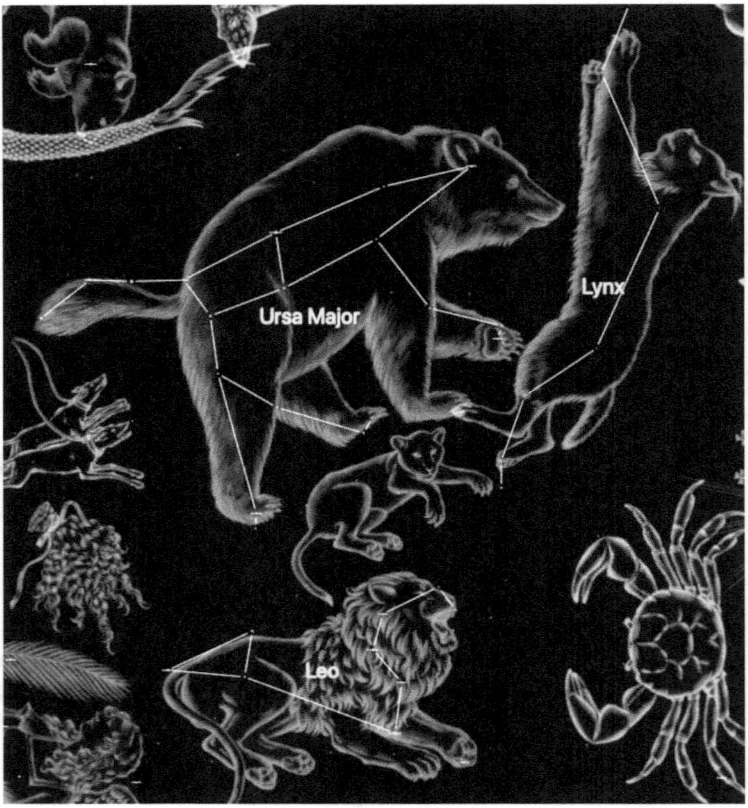

Figure 3.15 - Leo/Ursa/Lynx cluster.
Image courtesy Starry Night Astronomy.

From **Virgo**'s immaculate dawn to **Libra**'s final reckoning, the constellations mark history's complete arc. Unlike fleeting human records the heavens proclaiming “the glory of God” to all peoples regardless of epoch or iteration. Scripture, not culture, anchors their meaning ensuring consistency cross cycles inviting us to see history's repetition. through God's unchanging story, written in starlight.

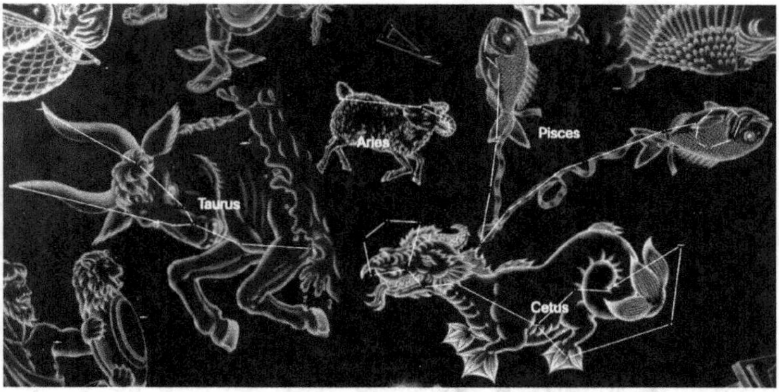

Figure 3.16 - The Christian Era
Taurus/Aries/Pisces (Cetus) cluster.
Image courtesy Starry Night Astronomy.

Figure 3.17 - Judgement through Redemption cluster
Aquarius/Capricorn/Sagittarius/Clouds of Heaven/Scorpio/Libra
Image courtesy of Starry Night Astronomy

3.6 Cancer, Gemini and Off-the-Ecliptic

Not everything in the night sky fits neatly into the cycle we have been tracing. This is worth pausing on—not because it undermines the argument, but because it actually illuminates something important about how God's celestial narrative works. The twelve-sign zodiac includes every constellation along the ecliptic, but not all of them play active roles in the biblical cycle as Daniel's vision describes it. Two of them—Cancer and Gemini—are within the ecliptic belt but absent from the cyclical narrative in any discernible way. Others lie off the ecliptic entirely, outside the zodiacal band, yet occupy positions in the sky that are clearly visible and clearly meaningful to anyone who looks carefully.

These gaps and extras are not flaws in the system. They are features of it. They suggest that the celestial narrative God embedded in the stars is richer and more complex than any single telling can encompass—that the twelve-constellation cycle this book has traced is one thread in a much larger tapestry.

There are innumerable other constellations and images in the night sky. In other cycles these constellations may play a different roll in illustrating history of that cycle. In other cycles Perseus, Hercules, Orion, Andromeda, Pegasus; as well as, Cancer and Gemini may have played a defining roll in the narrative of another version of Daniel and John.

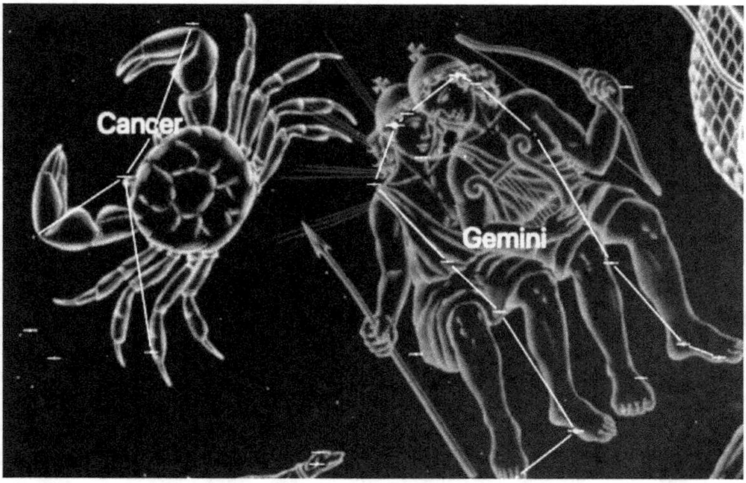

Figure 3.18 - Cancer and Gemini.
Image courtesy Starry Night Astronomy.

Cancer - occupies the zodiacal position between Gemini and Leo—nestled in a region of sky that is, by astronomical standards, relatively faint and unremarkable. It contains no bright stars. Its most notable feature is a star cluster called the Beehive (Praesepe), a dense gathering of stars that looks, under good conditions, like a small smudge of light to the naked eye. Ancient astronomers found this cluster mysterious and gave it ominous significance

—in some traditions, it was considered a 'manger' or a 'gateway.'

In the biblical cyclical narrative, Cancer does not play an assigned role in Daniel's sequence of empires, nor in the Revelation sequence of judgment and renewal. It sits, in a sense, between cycles—positioned between Leo (Babylon, the beginning of Daniel's vision) and Gemini (which we will address in a moment). Its absence from the active narrative may itself be meaningful: like the rests between musical phrases, the constellations that do not 'speak' within a given cycle may mark the transitions between the audible movements of the cosmic symphony.

Historically - Cancer was known in Babylonian astronomy as 'the Crab' or 'the Crayfish.' In Greek mythology it was the crab sent by Hera to distract Heracles during his battle with the Lernaean Hydra—stepped on and crushed in the process, then immortalized as a constellation in consolation. The somewhat anticlimactic nature of Cancer's myth reflects its astronomical position: a faint, transitional sign, present but not dominant, marking a threshold rather than an era.

Gemini - is one of the most immediately recognizable constellations in the northern sky—two bright stars (Castor and Pollux) marking the heads of the twins, with fainter stars trailing below them in parallel columns. Its position in the ecliptic places it within the crossing of the Milky Way—what this book identifies as the River Ulai, the first heavenly crossing—meaning that Gemini sits right at the boundary between two major sections of the cyclical narrative.

The duality of Gemini—two figures, inseparable yet distinct—invites a rich range of theological associations without committing to any single one with certainty. One might imagine the divine duality of YHWH and Elohim—the God who is both transcendent and near. Or, the theanthropic (God-man) nature of Christ Jesus, fully divine and fully human. Or marriage, as the duality of two becoming one flesh. Or the relationship between the old and new covenants, the Hebrew Bible and the New Testament, inseparably bound yet distinct in form.

Historically - Among the oldest recognized star groupings in human history, Gemini appears in Babylonian astronomical tablets as—"The Great Twins"—identified as Meshlamtaea and Lugalirra, twin deities guarding the entrance to the underworld. During the third millennium BC, the sun rose in Gemini at the summer solstice, making the twins celestial gatekeepers between light and darkness. Egyptian astronomers recorded the pair as two sprouting plants, symbols of duality woven into creation. The Greeks renamed them Castor and Pollux — the Dioscuri — sons of Zeus, patrons of sailors, embodiments of mortal and immortal held together in permanent union.

We are cautioned against carrying these associations too far. Passages like Isaiah 47:13-14 and Deuteronomy 4:19 warn against the kind of astrological interpretation that treats constellations like Gemini as having influence over human destiny. The ancient Israelites were aware of the constellations—as the many references in the wisdom literature confirm—but their focus was on God's sovereignty, not on the constellations' own power. Daniel's concern was astronomical and theological: signs to be read and interpreted for the cycles they describe in the broadest sense. Not every sign is active in every cycle.

Off the Ecliptic: When Other Cycles Call Other Stars

Beyond the ecliptic zodiac—the twelve constellations that line the sun's apparent path through the sky—there are dozens of other constellations that occupy the wider dome of the night sky. Ursa Major and Ursa Minor circle the north celestial pole. Orion strides across the celestial equator. Perseus raises his sword above the horizon. Andromeda stretches her arms across the sky. Cygnus the Swan flies along the Milky Way. Draco the Dragon coils around the pole. Hercules kneels in the northern sky. Lyra holds its bright Vega near the zenith on summer nights.

These constellations—and many others—are part of God's celestial canvas but do not figure in the specific cycle that Daniel and Revelation describe. Their absence from the biblical narrative is not evidence of their meaninglessness. It is evidence that the narrative we have been tracing is one cycle among many—and that other cycles may have used, or will use, other constellations to mark their own history.

Consider what this implies. If the cycle we have been describing spans something in the range of 12,000-15,000 years, and if the Precession of the Equinoxes provides a larger astronomical clock that ticks through all twelve zodiacal signs in approximately 25,920 years, then there have been cycles—and will be cycles—whose histories are marked by constellations that this cycle's narrative never touches. The people living in those cycles, guided by the same God who gave Daniel his visions and John his Revelation, will be given the eyes to see the signs written for their own age. The symbols may differ. The story they tell—creation, fall, judgment, redemption, renewal—will be the same.

Imagine what those other cycles might have used. Orion—the great hunter, striding across the sky—might have served as a marker for an era of violent conquest and great personal ambition. Perseus—the hero who rescued Andromeda from the sea monster—might have described a cycle's redemptive turning point in imagery that resonated with a culture whose stories ran differently from ours. Draco the Dragon, coiling around the celestial pole, might have marked the presence of the adversary in a cycle whose astronomers watched a different part of the sky for that signal. Cygnus the Swan, flying along the Milky Way, might have served as the 'clouds of heaven' imagery for a cycle whose return of the Messiah was heralded by the flying bird rather than the rider on a horse. Lyra's bright Vega, once the pole star (due to precession) and again the pole star in approximately 12,000 years, might anchor cycles whose north star tells a different navigational story.

None of this is speculation about what those cycles contained. It is speculation about how God might have written their story in the sky. The specific content of those cycles—the empires that rose, the prophets who spoke, the remnant that survived—only the people of those cycles could have known. What we can say is that the same God who placed the stars as signs for our cycle did so for every cycle before and after. The celestial vocabulary is his. Each generation—in each cycle—receives the portion of it they need.

All cycles begin and end fundamentally the same: with judgment, redemption, and renewal. The specific constellation images may vary. The story they tell does not. Virgo to Libra, creation to justice, beginning to end—this is the shape of God's recurring purpose in the cosmos, written in light that outlasts every civilization that has ever looked up at it and tried to understand what it was saying.

We are not the first to look and we will likely not be the last. But we may be among the first in a very long time to be given the eyes to read what was always there—exactly as Daniel 12:4 promised: 'Many will range far and wide and *the* knowledge will increase.'

PART 4: A Broader View

Romans 8:20-21
*20 For the creation was subjected to futility, not willingly,
but because of Him who subjected it in hope;
21 because the creation itself also will be delivered from
the bondage of corruption into the glorious liberty of the
children of God.*

4.1 Sir Isaac Newton and the Stars

Refer to Isaac Newton's formative work on prophecy, Observations on the Prophecies of Daniel and the Apocalypse of St. John (*Prophecies*), published in 1733.

Daniel is most distinct in order of time, and easiest to be understood, and therefore in those things which relate to the last times, he must be made the key to the rest.
(Prophecies, p. 14)

Daniel was in the greatest credit amongst the Jews, till the reign of the Roman Emperor Hadrian. And to reject his prophecies, is to reject the Christian religion. For this religion is founded upon his prophecy concerning the Messiah.
(Prophecies, p. 24)

Sir Isaac Newton (1642-1727), a towering figure of the Enlightenment, is celebrated for his laws of motion and universal gravitation, which reshaped our understanding of the cosmos. Yet beneath his scientific brilliance lay a lesser-known passion—a relentless quest to decode the mysteries of scripture and history through the lens of celestial patterns. In his posthumously published work of 1728, *The Chronology of Ancient Kingdoms Amended*, he turned his analytical mind to ancient timelines, convinced that the Bible held a precise historical framework, synchronized with the stars. To be sure, Newton was not concerned in the least with constellations, as pursued in this book, but was convinced that he could coordinate specific events in history to specific alignments of stars.

Born in Woolsthorpe, England, on December 25, 1642 AD, Newton's life straddled a world in transition. The Scientific Revolution was dawning, yet medieval mysticism still lingered. A prodigy at Cambridge, he formulated calculus and optics by his mid-twenties, but his private pursuits were equally audacious. Newton amassed a library of over 1,700 books, many theological, and penned millions of words on alchemy, Biblical exegesis and chronology —works he kept largely hidden, fearing ridicule or heresy charges.

In *The Chronology*, Newton sought to revise the chaotic timelines of antiquity—Egyptian, Greek, and Biblical—using astronomical data and scriptural clues. He argued that traditional dates, bloated by myth and pride, needed pruning. As in all his pursuits, his methods were meticulous, blending Kepler's star charts with his own observations, yet it was driven by a theological conviction: God's creation was orderly, and the heavens bore witness to His plan.

Newton's heresies reached beyond his theological view of the stars as God's cosmological clock, he was also a very ardent anti-Trinitarian. As such, he was in direct opposition to the Catholic Church's position on the Trinity and continually trying to hide his theological research and conclusions. Additionally, he was quite vocal in his theological papers about staying away from Apocalyptic "date-setting", so it may come as a bit of a surprise that he did, in fact on numerous occasions, do that very thing. Without going into the tedium of his mathematics (basic math, not anything terribly sophisticated) he attempted to set a few dates for the beginning of the Kingdom of God on earth. Most dramatically was one that, in his day, was in the unimaginable future, but is now a date we can foresee as right on our doorstep - **2060 AD**.

For a thorough review of the documents and research that brought us this revelation, see Statement on the Date 2060, by Stephen D. Snobelen, Professor of Humanities, History of Science and Technology, University of King's College; Halifax, Nova Scotia, Canada, at https://isaac-newton.org.

4.2 Summary: What it all Means

If we are to understand scripture and the world around us through the lens of recurring cycles, cycles of creation, judgment, and renewal—cycles that transcend linear history—it is helpful to recognize cyclical loops within linear history. These loops are the patterns that Santayana urged us to recognize. The physical world and cosmos at large, human history and sociology, and even ancient cultural memory bear witness to such patterned loops. Such are the repeating motifs and rhythms that are occurring around us and within us that we may miss because we have become so accustomed to their presence that we are no longer aware of them. Or, they are so expansive that they make us feel powerless—therefore, we ignore them.

Consider first the *Precession of the Equinoxes*, a slow, 25,920-year cycle caused by the wobble of the earth's axis. Ancient civilizations such as the Egyptians, Hindus, and Maya not only recognized this phenomenon but incorporated it into their cosmologies. The zodiacal periods, roughly 2,160-year epochs tied to each of the 12 primary zodiacal images and, together, tied to this precession, coincide with major cultural and spiritual shifts in human history. The Age of Taurus aligned with bull-worshiping cultures (Egypt, Mesopotamia); the Age of Aries with militaristic expansion and sacrifice; the Age of Pisces with the Christian era. Each age offers symbolic resonance with Biblical stages of human development, and each appears to mark transitions within God's redemptive cycle.

Likewise, *economic cycles*, like the Kondratiev Wave or business cycles, demonstrate repeating booms and busts that span decades, often corresponding with societal changes, wars, or innovations. *Social cycles*, such as those observed by philosophers like Ibn Khaldun or Oswald Spengler, describe predictable phases of societal rise, corruption, collapse, and renewal.

Weather cycles, including major glaciation events or El Niño patterns, remind us that climate is not stable but subject to grand oscillations. Even the *lunar and solar cycles*, foundational to agricultural societies and embedded in the calendars of agricultural societies and eras, underscore the repetitive structure of God's design. These are not random but ordered rhythms, loops to remind us that all things move according to patterns.

When read alongside scripture, these loops support the notion that history, though measured and chronicled as a straight line progression, is a tapestry of repeated themes, a spiraling ascent toward ultimate restoration. Just as Ecclesiastes 1:9 observes, “That which has been is what will be, that which is done is what will be done, and there is nothing new under the sun,” (NKJV) so too does the Bible’s cosmic scope affirm a God who reveals Himself again and again, in every generation, through repeating signs.

What’s more, remnants of prior civilizations suggest that we are not the first iteration of this pattern. Throughout this book, we have traced the Biblical cycle from Genesis to Revelation and back to Genesis, seeing our generation as one within a long arc of divine storytelling. But we are not the first to walk among the ruins of a forgotten world.

Consider the *Caral*, in ancient Peru, predating the Inca by thousands of years and contemporaneous with early Egypt and Mesopotamia. Caral's advanced architecture and organized urban planning raise difficult questions: who taught them, and from where did their knowledge come? Similarly, the *Toltec* civilization of Mesoamerica, shrouded in myth and predating the Aztecs, is remembered as the bringers of order and wisdom, almost god-like in their legacy. These cultures seem to have inherited knowledge from an even older source, one now lost to time.

The *Mayan long-count calendar*, ending one major cycle in 2012, reflects an advanced understanding of celestial mechanics and cyclic time. It didn't predict the end of the world, but the end of an age. This interpretation mirrors the Biblical notion of ages and generations, each part of a repeating cosmic narrative. The *pre-Egyptians* of the Nile Valley, often referred to as the builders of megaliths before the pharaohs, remain an enigma. The sophistication of their stonework and astronomical alignments suggest they were the inheritors, or survivors, of a previous iteration.

These ancient cultures, often disconnected by geography and language, share a recurring theme —they seem to remember something. Whether in their mythology, architecture, or calendars, they whisper of a time before time, of teachers and gods, of floods and renewals. These are not mere legends; they are cultural fossils, memories of past cycles, past falls, and perhaps even past redemptions. Scripture does not ask us to dismiss these echoes but to see them in context—as remnants of earlier iterations in God's great cycle of judgment and renewal.

What we find, then, is that the Bible not only accommodates this view of cyclical history, it affirms it. Creation and re-creation, judgment and mercy, exile and return, these are not isolated events but patterns. As this book has explored, these patterns are written in the stars and codified in Scripture itself. They converge in our generation, calling us to recognize the signs, discern the time, and return to the Shekinah presence of God. We are not simply standing on the shoulders of our ancestors—we are walking their path, repeating their decisions, and awaiting the same Savior who has delivered across every age.

4.3 God's Sovereignty and Will

Ecclesiastes 3:14-15 (JPS):

> *14I realized, too, that whatever God has brought to pass will recur evermore:*
>
> *Nothing can be added to it*
> *And nothing taken from it -*
>
> *and God has brought to pass that men revere Him.*
>
> *15**What is occurring occurred long since, and what is to occur occurred long since: and God seeks the pursued.***

Isaiah 46:8-11 (JPS):

> *8 "Keep this in mind, and stand firm!*
> *Take this to heart, you sinners!*
> *9 Bear in mind what happened of old;*
> *For I am God, and there is none else,*
> *I am divine, and there is none like Me.*
> *10 **I foretell the end from the beginning,***
> ***And from the start, things that had not occurred.***
> *I say: My plan shall be fulfilled;*
> *I will do all I have purposed.*
> *11 I summoned that swooping bird from the East;*
> *From a distant land, the man of My purpose.*
> *I have spoken, so I will bring it to pass;*
> *I have designed it, so I will complete it.*

In light of all this, we return full circle, just as Scripture itself does, not merely to the beginning, but to the pattern beneath beginnings. What Genesis unveils and Revelation reaffirms is not a tale told once, but a cycle layered across epochs. The presence of light before the sun, of vegetation before the luminaries, of Adam before gendered reproduction, these are not curiosities to be dismissed or harmonized with modern science. They are deliberate theological cues, pointing to a deeper metaphysical truth: that God's order transcends what we call natural law, and His glory sustains life directly, before and beyond creation's mechanisms.

Marriage and reproduction, far from being incidental developments in Eden, are shown to be purposeful additions following the foreknown fall. Adam, initially asexually created and lacking a mate, becomes male only in the presence of Eve, whose arrival introduces the possibility of "one flesh." Reproduction, then, becomes the way humanity persists after separation from God. Yet, in the resurrection, Jesus affirms that this function ends. We become "like the angels", not in power or location, but in mode of being. No longer subject to decay or driven by the necessity of lineage, we return to the original design: creatures who derive life not through biology, but through proximity to God.

This insight reframes the Genesis narrative not as a history of what was, but as a map of what will be. Eden is not merely a paradise lost, but a prototype of the end, one that Revelation magnifies and restores. In the new Jerusalem, where the Lamb is the light, where the river flows from the throne of God, and where the tree of life bears fruit monthly, we no longer need sun, moon, or even the institution of marriage. These things served a purpose in the fallen world, but in the cycle's final arc, they give way to something higher—direct communion with our Creator.

Understanding this cyclical vision changes how we interpret Scripture and how we live. We are not only moving forward through time; we are spiraling upward through divine recurrence. Creation, fall, redemption, and restoration are not isolated events, they are archetypes, unfolding again and again at varying scales. The end is not simply a return to the beginning, it is the fulfillment of the beginning's promise. The bride is not merely woman restored to man, but humanity restored to God.

Therefore, we stand not at the edge of linear time, but in the flow of God's eternal rhythm. We must ask with renewed urgency: are we nearing the conclusion of a cycle, and the birth of another? The signs in the heavens still speak. The Bride still prepares herself. The invitation still rings out: "The Spirit and the bride say, 'Come'" (Revelation 22:17). The garden waits. The river flows. The light shines, not from the sun, but from the Lamb. And so the cycle turns once more.

The flexibility and layered meaning of the Hebrew word *b'ra-sheet* opens a gateway into a deeper understanding of Biblical cycles and creation itself. Rather than presenting a singular, absolute beginning, *b'ra-sheet* invites us to consider the possibility of multiple beginnings, repeating epochs that unfold across time, space, and divine intention. By rendering Genesis 1:1 as "When God began to create…" or even "In **A** beginning…," we shift away from linear thinking and open ourselves to the idea that Genesis describes not the first moment of existence, but the commencement of a new creative phase following a cataclysmic rupture.

This framework aligns seamlessly with the apocalyptic vision of Revelation. The "new heaven and new earth" of Revelation 21:1 do not appear in a vacuum, they arise from the passing away of the first, just as Genesis 1:2 presents a world that is already formless, void, and shrouded in darkness. What if Revelation describes the very destruction that precedes the Genesis re-creation? In this view, the Bible's structure is not strictly linear but cyclic.

Revelation and Genesis are not terminal ends, but transition points in a divine rhythm that stretches across eons.

This narrative arc, from Genesis' formless void to Revelation's New Jerusalem, reveals Earth's history as a divine rhythm, not a linear tale. Each cycle begins with God's creative act, repairing what was broken, and concludes with renewal, only to begin anew. *B'ra-sheet*'s flexibility invites us to see creation as iterative, with Revelation's end flowing into Genesis' beginning. The stars, as eternal witnesses, guide us through these cycles, urging reflection—are we in the final turning? As we ponder, the constellations whisper God's timeless story, calling us to trust in His renewing purpose.

The stars, ordained as *signs* (Genesis 1:14), were placed with intentionality, not only to mark seasons but to signal the unfolding of divine purposes. Their positions and constellations speak to a cosmic story, one that predates human history and reflects a divine intelligence echoing through the ages. Understanding *b'ra-sheet* as the introduction to one cycle among many transforms our reading of Scripture. It becomes a narrative of recurrence, judgment, and renewal. It beckons us to pay attention not only to what is written, but to the patterns, gaps, and implications that emerge when we read Genesis and Revelation together, as one continuous, eternal return from light to light.

The Book of Daniel, set against the Babylonian exile, unveils a cosmic framework that transcends its historical context, revealing history as a cyclical saga of divine creation, rebellion, judgment, and renewal. Daniel's visions in chapters 7 and 8, shifting to a personal first-person narrative, harness the "signs" of Genesis 1:14, celestial constellations, to map this rhythm. These chapters are not mere prophecies for ancient empires but celestial blueprints, echoed in Revelation, that chart recurring cycles across millennia, possibly spanning 12,000–15,000 years or more. The stars, as God's enduring markers, narrate a story older than human memory, linking Genesis' creation to Revelation's renewal.

In Daniel 7, the four beasts—lion (Leo), bear (Ursa), leopard (Lynx), and horned beast (Taurus the bull)—symbolize empires (Babylon, Medo-Persia, Greece, Rome) while foreshadowing a broader cosmic cycle. The lion's majesty reflects Babylon's dominance, the bear's lumbering strength captures Medo-Persia's expanse, the leopard's agility mirrors Greece's swift conquests, and the bull's iron teeth embody Rome's enduring might. These are not just historical; they recur in Revelation 13:2's composite beast, suggesting history spirals, it does not progress linearly. The “Son of Man” (Daniel 7:13) and the “Ancient of Days” (Daniel 7:9) point to the Messiah’s ultimate triumph, restarting the cycle with divine order.

Daniel 8 zooms into this cycle, focusing on the Ram (Aries), symbolizing the Christian era’s spread, and the male goat (Capricorn), a chimeric adversary blending deception and power. The goat’s assault on the Ram’s horns (Daniel 8:7) prefigures challenges to faith, like Antiochus IV’s desecration or the Antichrist’s tribulation (Revelation 13:5–7). The 2,300 “evenings and mornings” (Daniel 8:14) align with Revelation’s seven-year tribulation, signaling restoration after trial. Capricorn’s dual nature, goat and dragon, mirrors Satan’s serpent-like opposition (Revelation 12:9), and the cycle concludes with Libra’s justice (Daniel 5:25), and Virgo’s purity (Isaiah 7:14), heralding renewal.

Daniel’s influence extends beyond his era, evident in the Magi’s quest for the Messiah’s star (Matthew 2:1–2), likely inspired by Daniel’s teachings to Babylonian priests. This star, possibly God’s Shekinah glory, announced Christ’s birth, tying Daniel’s celestial insights to the Christian narrative. The constellations, from Virgo to Libra, encode this story, guiding humanity through cycles of fall and redemption. Revelation, as Daniel’s echo, amplifies this, with its sea-beast (Revelation 13:1) and Lamb (Revelation 5:6) mirroring Daniel's imagery, suggesting a unified narrative across scripture.

This cyclic view, rooted in *b'ra-sheet*'s "when God began to create" (Genesis 1:1) and Revelation's "new heaven and earth" (Revelation 21:1), posit Earth's history as iterative. Enigmatic relics like Göbekli Tepe may hint at prior cycles, lost to time. Daniel and Revelation, twin lenses, reveal history's spiral, where empires rise and fall like constellations traversing the sky. As we ponder our place in this cosmic dance, the stars urge trust in God's eternal purpose (Isaiah 46:10), guiding us toward a renewal where the Lamb's light shines forever, restarting the cycle anew.

The story traced through this work, from the visions of Daniel to the redemptive arc of Revelation, presents a remarkable truth: the heavens have always been more than a backdrop to human activity. They are a sacred script authored by the Creator, a visual testament to His purposes woven through time. The constellations, when read through the lens of Scripture rather than superstition, become chapters in an ongoing divine chronicle, one that spans from the first light of creation to the final reckoning of judgment and renewal.

This celestial narrative, rooted in Genesis 1:14, was never meant to serve the mechanics of astrology or the dictates of human fate. The Hebrew Bible consistently warns against such distortions (Deuteronomy 4:19, Isaiah 47:13–14), insisting instead on a reverent recognition of the stars as signs, not forces to be appeased or manipulated, but testimonies to God's dominion and design. The constellations, then, are not fortune-tellers; they are faithful witnesses, proclaiming the same message echoed by the prophets: history is cyclical, judgment is inevitable, and redemption is assured.

From Virgo's symbol of divine birth and new creation to Libra's scales of justice and reset, the heavens map not just time but theology. The Lion, the Bear, the Lynx, the Bull, the Ram, and the Sea Monster are not arbitrary figures, they represent both earthly empires and spiritual forces that rise and fall within God's sovereign timeline. Jesus, the Lamb, appears at the pivot point, His sacrifice the fulcrum of all ages. Every image, whether it be Sagittarius launching judgment across the clouds of heaven or Aquarius pouring out wrath, directs the viewer not inward, as astrology tends to do, but upward and forward, toward the consummation of God's plan.

This interpretive framework insists on a cosmic consistency. The heavens declare God's glory

(Psalm 19:1), not mankind's speculation. They are enduring, fixed, and cyclical, not chaotic or whimsical. They do not bend to the imaginations of each age, but instead serve as a divine mirror, reflecting the truths found in Scripture across millennia. In them, we are not asked to discover our personal fate, but to discern the unfolding pattern of history itself, creation, rebellion, judgment, redemption, and renewal.

The visions of Daniel saw this pattern clearly. His visions did not simply reinterpret Babylonian motifs; they redefined them within a Biblical worldview, one that understands history as the working out of God's purposes through the rise and fall of nations and the eternal reign of the Messiah. The stars were His symbols, but Scripture was His key.

This narrative, rooted in Genesis 1:1's "when God began to create" and Revelation's "new heaven and earth" (Revelation 21:1), suggests iterative history, with relics like the trilithon stones of Baalbek's Heliopolis, hinting at past cycles. Daniel's influence, guiding the Magi to Christ's star (Matthew 2:1–2), ties his celestial insights to the Messiah's advent, affirming the stars as divine signposts. Revelation amplifies Daniel's imagery, its Lamb and beast mirroring his visions, weaving a unified scriptural saga.

“The heavens,” as Psalm 19:1 proclaims, “declare the glory of God, the sky proclaims His handiwork”, their meanings anchored in scripture, not cultural whims. Daniel’s creative liberty, envisioning a leopard in Lynx’s faint void, reflects the Hebrew Bible’s dynamic engagement with the stars, unbound by later astronomical conventions. As we ponder our place in this cosmic rhythm, the question arises—are we approaching Libra’s concluding justice? The stars, enduring across epochs, urge trust in God’s sovereignty, guiding us from Virgo’s dawn to Sagittarius’ victory, until the final renewal, where the Lamb’s light shines eternally, heralding a new cycle.

Today, we are called to read the heavens anew, not with the eyes of astrologers, but with the discernment of those who understand that God wrote His promises in the skies as well as in the text. The constellations speak, and their language is Scripture. And their message is this: God's purposes remain unshaken, His covenant stands, and the heavens will continue to proclaim it until all is fulfilled.

If my proposition about the Bible telling the story of cycles of creation and history is correct, it brings resolution to some of the most difficult interpretive challenges that have long puzzled believers and skeptics alike. Rather than reading Scripture as a linear, one-time historical progression, this cyclical perspective allows us to understand God's redemptive plan as one that unfolds across repeated iterations of creation, fall, judgment, and renewal, each cycle providing new opportunities for humanity to respond to God's grace, and each revealing His consistent character and sovereign design.

This cyclic framework casts new light on scripture's most enigmatic passages, resolving tensions that linear readings struggle to address. Among the most persistent of those tensions is what scholars call the delay of the parousia [G3952]—the return or second coming of Christ—the apparent gap between the urgency of the New Testament's prophetic language and the two-thousand-year interval that has since elapsed. The New Testament writers do not speak with vague futurism. They speak as men who believe the consummation of history is within arm's reach of their own generation.

Paul writes to the Romans: "Now our salvation is nearer than when we first believed...The night is far spent, the day is at hand" (Romans 13:11–12). To the Corinthians he is even more direct: "the time is short" (1 Corinthians 7:29), and "the ends of the ages have come upon us" (1 Corinthians 10:11). To the Philippians: "The Lord is at hand" (Philippians 4:5). The writer of Hebrews urges his readers to keep meeting together "all the more as you see the Day approaching" (Hebrews 10:25), adding moments later: "For yet a little while, and He who is coming will come and will not tarry" (Hebrews 10:37). James adds his own urgency: "the coming of the Lord is at hand...the Judge is standing at the door" (James 5:8–9). John writes plainly in his first epistle: "Little children, it is the last hour" (1 John 2:18). And Paul, in what may be the most striking imminence statement in all of the epistles, writes of "we who are alive and remain until the coming of the Lord"—placing himself and his readers among those who expect to witness the event personally (1 Thessalonians 4:15).

Then there is Jesus Himself. In Matthew 16:28 He declares: "Assuredly, I say to you, there are some standing here who shall not taste death till they see the Son of Man coming in His kingdom." And in the Olivet Discourse—His most sustained prophetic address—He states without qualification: "this generation will not pass away till all these things take place" (Matthew 24:34). The Greek word rendered generation here is genea [G1074], a term used consistently in the New Testament to refer to the contemporaries of the speaker. For the disciples hearing these words, the first and most natural meaning was unmistakable: the consummation was not centuries away. It was imminent, and they were to live in the full expectation of it.

And yet here we stand, two thousand years later.

Consider Revelation 1:1, where John speaks of "things which must shortly take place." For centuries, believers have grappled with the apparent delay of Christ's return, yet within the cyclical framework, "shortly" signals the swift culmination of an epoch, not an imminent calendar date. The tribulation (Capricorn, Daniel 8:14: 2,300 days), Christ's return (Sagittarius, Revelation 19:11), and the "new heaven and earth" (Revelation 21:1) unfold rapidly in God's eternal perspective, where "one day is as a thousand years" (2 Peter 3:8). It is worth noting that Peter introduces that statement in the precise context of defending the Lord against the charge of tardiness—"scoffers will come in the last days...saying, 'Where is the promise of His coming?'" (2 Peter 3:3–4)—suggesting the earliest believers were already confronting the very tension this framework resolves.

The cyclic model does not ask us to set aside the plain, immediate meaning of Jesus's words in Matthew 24:34. He was speaking directly to those standing before Him, and they were right to take His words personally and urgently. The first-century church lived on the forward edge of something that genuinely could have culminated in their lifetimes—and indeed, for those believers who witnessed the destruction of Jerusalem in 70 AD, the weight of His warnings proved devastatingly real. But Scripture consistently reveals that its deepest prophecies carry more than one layer of fulfillment. Jesus spoke both to His immediate audience and across the full span of the covenant age He was inaugurating. In the cyclic model, genea carries this dual weight naturally: it names the generation standing before Him in first-century Judea, and simultaneously encompasses the full covenant generation—all those who belong to the current age, the era of "fishers of men" (Matthew 4:19), stretching from Pentecost to the final harvest and culminating in Libra's concluding justice (Daniel 5:25ff). Every generation within that arc has lived rightly in the expectation of His return, because every generation stands within the same cycle, and any of them could have seen its end. None were wrong to watch. None were wrong to wait.

"Behold, I am coming quickly" (Revelation 22:7, 12, 20) is therefore not a failed promise to the first century, nor merely a distant comfort to the last. It is a covenantal certainty addressed to all who live within the cycle—a declaration that when the appointed moment arrives, it will come swiftly, suddenly, and without further delay. In the reckoning of the One for whom "a thousand years are like a watch in the night" (Psalm 90:4), the entire span of Christian history is but the closing hours of a single long day. "Shortly" was always the right word.

For Christians, this perspective deepens faith, showing God's grace, renewing humanity in each cycle, as Aries' Lamb (John 1:29) overcomes Capricorn's beast (Daniel 8:5). Skeptics may find a bridge between scripture and archaeology, where megaliths hint at lost cycles. Astronomers may marvel at the Mazzaroth's divine design, predating Ptolemy's astrology. Psalm 19:1 proclaims, "The heavens declare the glory of God," a truth etched in starlight across millennia and bridging innumerable cycles. As we stand in this cycle—perhaps the last—let us heed Philippians 4:7, seeking "the peace of God" and Jesus, "the light of men" (John 1:4), who guides us through history's spiral to eternity's embrace.

This cyclical view also provides important theological context for some of Scripture's most mysterious moments, such as the presence of evil *before* the fall of man. In Genesis 3, the serpent is already in the Garden, fully capable of temptation. In Genesis 6, we find the Nephilim, "the fallen ones," a result of spiritual corruption already at work. These episodes raise a pressing question: where did evil come from *before* the fall? If this is the first and only creation, the presence of such malevolence appears abrupt and inexplicable.

But if we are instead living in a recurring cycle, one preceded by prior ages, earlier judgments, and earlier cosmic rebellions, then evil's presence is more easily situated within a longer, repeating battle between light and darkness. Satan's final release in Revelation 20:7–10, long after a thousand years of peace, fits this pattern. It is not random but cyclical, an echo of previous rebellions and a prelude to final restoration.

Perhaps most helpfully, this framework resolves one of the most divisive debates in modern theology and science: young earth or old earth? If the Biblical narrative focuses on *this current cycle*, a phase of history perhaps just a few thousand years old, then the young earth timeline makes sense *within the context of our redemptive story*. While, at the same time, acknowledging cycles allows us to affirm the scientific evidence for an earth and universe that are incomprehensibly old. The Genesis account, then, is not a record of the universe's *first* beginning, but the most recent chapter in a much older divine symphony.

This dual truth, that the earth is both ancient and yet young in its redemptive cycle, honors both the testimony of Scripture and the insights of science. The Bible does not need to contradict observable evidence; it needs only to be read in the framework it provides, one of recurring cycles, where every generation is given its own moment of decision, and every iteration points us back to the same Savior. Each iteration starts and ends basically the same.

In the end, this view reinforces the great message of Scripture: God is not slow, inconsistent, or distant. He is patient, just, and ever-present. His story is always unfolding, and we are invited, right now, in this generation, to see, to respond and to believe.

4.4 On the Integrity of the Bible

Throughout this book, I have endeavored to show how the Old and New Testaments, when understood together, form a single, cohesive narrative. They are not two disconnected books or opposing dispensations, but two halves of the same divine testimony, a seamless revelation of God's unfolding purpose through time. This unified story is not merely about ancient peoples or past empires. It is about us, the people of this generation, this iteration of the ongoing cycle of history revealed in the stars and echoed in the pages of Scripture.

From Genesis to Revelation, the Bible reveals a consistent pattern: creation, fall, redemption, and renewal. Genesis begins with the origins of the heavens and the earth, the forming of humankind, and the fracturing of that creation through sin. Revelation ends with a new heaven and new earth, the restoration of all things, and the final victory over evil. These are not separate stories, they are the bookends of the current epic, with every prophet, psalm, gospel, and epistle contributing to its fullness.

It is common today for readers to approach the Bible with suspicion, assuming contradictions, errors, or inconsistencies. Many view the Old

Testament as obsolete or harsh, and the New Testament as either a total replacement or a new invention. This is a misunderstanding. What appear to be contradictions are often just uncertainties that dissolve under careful grammatical and contextual study. A deeper examination reveals harmony, not only in theology but also in vocabulary, symbolism, and prophetic continuity. It is why I've taken care in this work to provide quick access to grammatical guidance and word studies, especially where English translations obscure the richness of the original Hebrew or Greek.

The Bible is not a puzzle to be solved by casual reading. It is a layered and living text, and like any great narrative, its meaning becomes clearer the more we engage with it. Scripture itself affirms this: "It is the glory of God to conceal a matter, but the glory of kings is to search out a matter" (Proverbs 25:2). The Bible rewards deep study because its Author is consistent, purposeful, and sovereign across every page.

To rightly understand what the Bible means, you must first understand what it says. Too many interpretations are built on isolated verses, misunderstood phrases, or assumptions inherited from culture rather than drawn from Scripture. That is why this book has labored to bring Biblical language, symbols, particularly those in the heavens, and overarching narrative themes into alignment. When read through this lens, the Bible does not contradict itself. Rather, it reveals a stunning continuity, a story told across centuries and millennia, across covenants, across the stars themselves.

This generation stands within that story. Our lives are part of the same cycle witnessed by Daniel, anticipated by Isaiah, fulfilled in Christ, and foretold by Revelation. The Bible speaks to all times, but it especially speaks to this time. It tells one story, God's story, and we are living in its next chapter.

4.5 Final Thoughts

The world in which we live is undeniably broken. It is fractured by sin, marked by injustice, plagued with suffering, and groaning for redemption. This condition is not simply a matter of political systems or human progress, it is a spiritual crisis, a fallenness that has touched every heart, every generation, every corner of creation, yet this world was not always so. There was once harmony, beauty, and divine presence—the Shekinah, the manifest presence and glory of God dwelling among His people.

The great question of every generation has been how to return to that Presence. Philosophies, empires, and religions have each offered their own paths, but none can restore what was lost. Only one way has been revealed by God Himself, the way of salvation through the Lord Jesus Christ. He is not one among many options; He is the fulfillment of the story told in the stars and written in the Word. From the beginning in Genesis to the consummation in Revelation, all of Scripture points to Him.

Like the Magi who followed the star, driven by faith and anticipation, we too are called to seek Him with diligence and humility. That search is not fruitless. For those who seek Christ, He promises to be found. In Him, we are restored to the Shekinah, the indwelling presence of God, and we receive what the world cannot offer: “the peace of God, which surpasses all understanding” (Philippians 4:7).

No matter where you stand today, hear this final invitation clearly: seek the Savior, trust in Him, and walk in the light of the One who writes His promises across the heavens.

Bibliography & Reference Materials by book Part

GENERAL REFERENCE MATERIALS

(1) The New King James Version (NKJV). Thomas Nelson Press 1982.

(2) JPS Hebrew-English TaNaKh (JPS). Jewish Publication Services 2000.

(3) The English Standard Version (ESV). Crossway 2008.

(4) Strong, Dr. James. The Strong's Exhaustive Concordance of the Bible. Zondervan 2001.

(5) Gesenius' Hebrew and Chaldee Lexicon. Baker Book House 1979 (1st edition by Samuel Baxter & Sons 1847).

(6) Vine's Complete Expository Dictionary of Old and New Testament Words, Thomas Nelson Press 1984, 1996.

(7) Thayer, Joseph H. Greek-English Lexicon of the New Testament, Hendrickson Publishing 1896. (Coded with Strong's Concordance Numbers)

(8) Brown, F.-Driver, S.-Briggs, C. Hebrew and English Lexicon, Hendrickson Publishers 1906. (Coded with Strong's Concordance Numbers)

(9) 1 Enoch, especially ch. 91 & 93; Apocrypha, numerous sources.

(10)2 Baruch 12; Apocrypha, numerous sources. Part 1: An Old & New View

PART 1: AN OLD & NEW VIEW

(11)Foundation for the Study of Cycles, website cycles.org.

(12)Carroll, Sean. Dark Matter, Dark Energy: The Dark Side of the Universe; published by The Great Course, copyright The Teaching Company 2007.

(13)Clegg, Brian. Dark Matter and Dark Energy: The Hidden 95% of the Universe. Hot Science / Icon Books 2019.

(14)Einstein, Dr. Albert. Ideas and Opinions. Bonanza Books 1964.

(15)Einstein, Dr. Albert. Relativity: The Special and the General Theory. Bonanza Books 1961.

(16)Hertog, Thomas. On the Origin of Time: Stephen Hawking's Final Theory. Bantam Books 2023.

(17)Hawking, Dr. Stephen. A Brief History of Time. Bantam Books 1998.

(18)Boulter, Dr. Carmen. The Pyramid Code video series. Magdalena Productions 2009.

(19)Hancock, Graham. Magicians of the Gods. Coronet 2015.

(20) Collins, Andrew. Gobekli Tepe: Genesis of the Gods. Bear & Company 2014.

PART 2: GENESIS AND REVELATION

(21)Kidner, Derek. Kidner Classic Commentaries: Genesis, Tyndale/InterVarsity Press 1967.

(22)Morris, Dr Henry. The Beginning of the World: A Scientific Study of Genesis 1-11. Master Books 2024.

(23)Howell, Donna & Horn, Dr. Thomas. Before Genesis. Defender Publishing 2024.

(24)Howell, Donna & Horn, Dr. Thomas. Before Genesis: The Unauthorized History of Tohu, Boho, and the Chaos Dragon in the Land Before Time. Defender Publishing 2023.

(25)Bauval, Robert & Brophy PhD, Thomas. Black Genesis: The Prehistoric Origins of Ancient Egypt. Bear & Company 2011.

(26)Wilson, Dr. Josh. Answers Research Journal 2018. Linguistic Traits of Hebrew Relator Nouns and Their Implications for Translating Genesis 1:1.

(27)Wilson, Dr. Josh. Answers Research Journal 2018. Syntactical Features of Hebrew Genitive Clauses and Their Implications for Translating Genesis 1:1.

(28)Answers in Genesis (AiG) website can be found at https://answersingenesis.org and is free for public access.

(29)Oard, Michael. Frozen in Time: Woolly Mammoths, The Ice Age, and the Biblical Key to Their Secrets. Master Books 2004.

(30)Petrovich, Dr. Douglas. Patterns of Evidence: Hebrew the World's Oldest Alphabet video series. Thinking Man Films 2020.

(31)Petrovich, Dr. Douglas. Patterns of Evidence: The Moses Controversy video series. Thinking Man Films 2019.

(32)Petrovich, Douglas. The World's Oldest Alphabet: Hebrew as the Language of the Proto-Consonantal Script.Jerusalem: Carta, 2016.

(33)Rohl, Dr. David. From Eden to Exile: The 5000-year History of the People of the Bible. Greenleaf Press 2009.

(34)Bright, John. A History of Israel. 4th ed. Louisville: Westminster John Knox Press, 2000.

(35)Tov, Emanuel. Scribal Practices and Approaches Reflected in the Texts Found in the Judean Desert. Studies on the Texts of the Desert of Judah, vol. 54. Leiden: Brill, 2004.

(36)Szulc, Tad, (2001, December). Journey of Faith. National Geographic, Vol. 200(6), pp. 90-129.

(37)Jewish Encyclopedia. "Satan." In The Jewish Encyclopedia: A Descriptive Record of the History, Religion, Literature, and Customs of the Jewish People, vol. 11, edited by Isidore Singer. New York: Funk & Wagnalls, 1905.

(38)Jews for Judaism. "The Jewish View of Satan." Jews for Judaism (website), accessed 2026. Los Angeles: Jews for Judaism International.)

(39)My Jewish Learning. "Do Jews Believe in Satan?" My Jewish Learning (website), accessed 2026. New York: 70 Faces Media.

(40)Walton, John H. Ancient Near Eastern Thought and the Old Testament: Introducing the Conceptual World of the Hebrew Bible. 2nd ed. Grand Rapids: Baker Academic, 2018.

PART 3: DANIEL, IN THE NEXUS OF TIME

(41)Astronomical images used throughout this text courtesy of Starry Night Astronomy, a Simulation Curriculum entity, www.starrynight.com.

(42)Missler, Chuck. The Book of Daniel: A Commentary, Koinonia House 2004.

(43)Missler, Chuck. The Book of Revelation: A Commentary, Koinonia House 2001. Lynn, Dr. Heather. Evil Archaeology. Disinformation Books 2019.

(44)Collins, John J. Daniel: A Commentary on the Book of Daniel. Hermeneia: A Critical and Historical Commentary on the Bible. Minneapolis: Fortress Press, 1993

(45)Anthony, David W. The Horse the Wheel and Language. Princeton University Press 2007.

(46)Spinney, Laura. Proto: How One Accident Language Went Global. Wiktor Quill Books 2025.

(47)Bullinger, E. W. The Witness of the Stars. London: 1893. Reprint, Grand Rapids: Kregel Publications, 1967.

(48)Seiss, Joseph A. The Gospel in the Stars. 1882. Reprint, Grand Rapids: Kregel Publications, 1972.

(49)Johnson, Jr, Robert Bowie. The Parthenon Code: Mankind's History in Marble. Solving Light Books, 2004.

(50)Bullfinch, Thomas. The Complete and Unabridged Bullfinch's Mythology. Random House, Inc. 1993; pp. 117,119-120,161, 248.

(51)Rochberg, Francesca. The Heavenly Writing: Divination, Horoscopy, and Astronomy in Mesopotamian Culture.Cambridge: Cambridge University Press, 2004.

(52)Boyce, Mary. A History of Zoroastrianism. 3 vols. Handbuch der Orientalistik, Series 1. Leiden: Brill, 1975–1991.

(53)Yamauchi, Edwin M. Persia and the Bible. Grand Rapids: Baker Book House, 1990.

(54)Hunger, Hermann, and David E. Pingree. MUL.APIN: An Astronomical Compendium in Cuneiform. Archiv für Orientforschung, Beiheft 24. Horn, Austria: F. Berger, 1989.

(55)Evans, James. The History and Practice of Ancient Astronomy. New York: Oxford University Press, 1998.

(56)Powell, Robert. History of the Zodiac. Sophia Academic Press 2007; pg 4-5, Appendix II: Newton's Chronology, p. 195ff.

(57)Skjaervo, Prods O. The Spirit of Zoroastrianism, Yale University Press 2011.

(58)Hinnells, John R. Editor. Zoroastrianism: Textual Sources for the Study of Religion. University of Chicago Press 1990.

PART 4: A BROADER VIEW

(59)Snobelen, Stephen D. Isaac Newton: Theology, Prophecy, Science and Religion - Statement on the Date 2060. Originally released online: March 2003; updated May 2003 and June 2003; updated March 2025. Go to https://isaac-newton.org/statement-on-the-date-2060.

(60)Schumpeter, J. A. (1939). Business Cycles: A Theoretical, Historical, and Statistical Analysis of the Capitalist Process. McGraw-Hill.

(61)Ibn Khaldun. The Muqaddimah: An Introduction to History (F. Rosenthal, Trans.). Princeton University Press 1967.

(62)Spengler, O. The Decline of the West (Abridged Edition, H. Stuart Hughes, Ed.). Oxford University Press 1991.

www.ingramcontent.com/pod-product-compliance
Lightning Source LLC
LaVergne TN
LVHW022155200926
848481LV00006B/1753
* 9 7 9 8 9 9 9 4 9 2 4 4 9 *